FOUNDATIONS AND APPLICATIONS OF DECISION THEORY

THE UNIVERSITY OF WESTERN ONTARIO SERIES IN PHILOSOPHY OF SCIENCE

A SERIES OF BOOKS

ON PHILOSOPHY OF SCIENCE, METHODOLOGY,

AND EPISTEMOLOGY

PUBLISHED IN CONNECTION WITH

THE UNIVERSITY OF WESTERN ONTARIO

PHILOSOPHY OF SCIENCE PROGRAMME

VOLUME 13

FOUNDATIONS AND APPLICATIONS OF DECISION THEORY

VOLUME II

EPISTEMIC AND SOCIAL APPLICATIONS

Edited by

C. A. HOOKER

The University of Western Ontario

J. J. LEACH

The University of Western Ontario

and

E. F. McCLENNEN

Washington University, St. Louis

D. REIDEL PUBLISHING COMPANY

DORDRECHT : HOLLAND / BOSTON : U.S.A.

Library of Congress Cataloging in Publication Data

Main entry under title:

Foundations and applications of decision theory.

(The University of Western Ontario series in philosophy of science ; v. 13)

Papers resulting from a workshop held at the University of Western Ontario in the spring of 1975.

Includes bibliographies and indexes.

Contents: v. 1. Theoretical foundations.–v. 2 Epistemic and social applications.

1. Decision-making–Congresses. I. Hooker, Clifford Alan. II. Leach, James J. III. McClennen, Edward Francis, 1936– IV. University of Western Ontario. The University of Western Ontario series in philosophy of science ; v. 13.

T57.95.F68 658.4'03 77-25329

ISBN 90-277-0861-4 (set)

ISBN 90-277-0842-8 (Volume I)

ISBN 90-277-0844-4 (Volume II)

Published by D. Reidel Publishing Company,
P.O. Box 17, Dordrecht, Holland

Sold and distributed in the U.S.A., Canada, and Mexico
by D. Reidel Publishing Company, Inc.
Lincoln Building, 160 Old Derby Street, Hingham,
Mass. 02043, U.S.A.

Printed in The Netherlands

CONTENTS

VOLUME II

CONTENTS

VOLUME I

PREFACE

1. Introduction

In the Spring of 1975 we held an international workshop on the Foundations and Application of Decision Theory at the University of Western Ontario. To help structure the workshop into ordered and manageable sessions we distributed the following statement of our goals to all invited participants. They in turn responded with useful revisions and suggested their own areas of interest. Since this procedure provided the eventual format of the sessions, we include it here as the most appropriate introduction to these collected papers resulting from the workshop. The reader can readily gauge the approximation to our mutual goals.

2. Statement of objectives and rationale

(Attached to this statement is a bibliography; names of persons cited in the statement and writing in this century will be found referenced in the bibliography – certain 'classics' aside.)

2.1. *Preamble*

We understand in the following the Theory of Decisions in a broader sense than is presently customary, construing it to embrace a general theory of decision-making, including social, political and economic theory and applications. Thus, we subsume the Theory of Games under the head of Decision Theory, regarding it as a particularly clearly formulated version of part of the general theory of decision-making. The Theory of Games began its modern formal development with the pioneering work of von Neumann and Morganstern and has since been developed into a sophisticated mathematical discipline, among others by Arrow, Harsanyi, Luce, Nash, Raiffa, Rapoport and Sen. In recent times, this approach to decision-making has been increasingly applied to practical aspects of life, e.g., to the analysis of political and military strategy, to political institutions and group behaviour. Of course, micro-economic theory has for a very much longer time employed an explicit decision-theoretic (i.e., cost-benefit) analysis of rational behaviour which, though often cast in somewhat different terms, is essentially that of the Theory of Games. The Theory of Games does not exhaust the content of Decision Theory proper because the former in its present development does not treat of such dimensions as the Theory of Rational Acceptance (or Theory of Rational Belief, i.e., the rational acceptance and change of beliefs), the Theory of Value Dynamics and the theory of individual decision-making against nature. The Theory of Games needs to be properly imbedded in this larger context in order for it to be properly evaluated and fruitfully applied. The totality, together with the collection of applications to economics, politics and many other areas, constitutes a general Theory of Decision proper.

At the present time no example of a general Theory of Decisions proper exists that is even modestly developed, although several dimensions have been developed quite far (see below for selected references). One of the major purposes of the colloquium is to encourage the development of this larger perspective through the bringing together of outstanding leaders in each of the individual dimensions to obtain the critical interplay of their ideas.

Moreover, many applications of the Theory of Games to various subject matters are as yet in their infancy (e.g., to anthropology, organizational behaviour), with each employing a distinct terminology and a general lack of perception of common connections to a theoretical base. Another major objective of the colloquium is to contribute to the communications among the various disciplinary fields on this issue and to help to clarify the development of conceptual categories, terminology and theory in relation to the abstract core theory.

Both of these aims may properly be described as inter-disciplinary. In addition, the colloquium had several more conventional research goals.

Many of the more recent applications, or attempts at application, of the Theory of Games to various disciplinary subjects are highly controversial. One aim of the colloquium is to extend and clarify these applications through exchange of ideas between experts in the disciplinary fields and those who have played a leading role in the development of the abstract theory.

In fact, the organisers of this colloquium have a particular intellectual perspective on the unification of applications to various fields and its intellectual significance. First is the importance of perceiving that there is a unification of theory to be achieved. And second, there is the establishment of a unique unification of explanatory theory with normative theory. Starting with a formal uninterpreted calculus of rational choice, derived from Game Theory, two types of parameters (descriptive-explanatory and normative) can be plugged into the model to yield both an explanatory social theory of why people believe and act as they do, and a normative theory of how they should act (moral) and believe (epistemic), given their initial beliefs and preferences. The novel generality of the theory, then, derives from the fact that it applies not only to the normative domains of ethics and epistemology. The main thrust of the thesis, and a major goal of the colloquium is to explore the view that the concept of rationality has both explanatory and normative force, that both dimensions of this concept can be theoretically integrated, that the explanatory force has far-reaching applications in all the social scientific disciplines, and that the normative theories must become empirically controlled by the results of the explanatory theory.

Finally, the attempt to extend the applications of the Theory of Games to various subject matters (as one of their theoretical foundations) has made clear certain fundamental limitations in the present development of the abstract theory itself – this has to do with the notion of interdependency in decision-making and the development of a dynamical theory. An important aim of the colloquium is to crystalize and coordinate the various research work devoted to the elaboration of the Theory of Games within the context of various interdependence assumptions.

In the light of these remarks the colloquium organizers have chosen to develop the colloquium proceedings in four basic sections. These are now discussed in more detail.

2.2. *Foundations of Decision and Game Theory*

Foundational issues have already been briefly reviewed in the preamble. As was noted, recent attempts to extend the traditional Theory of Games to actual social and political situations have made certain limitations in that theory increasingly apparent. Specifically, virtually all the original foundational work done in Game Theory (as a theory of interdependent decision-making) has been predicated on the assumption that such a theory is constructable as an extension of the theory of individual, independent decision-making against nature (Statistical Decision Theory). Thus it has been assumed that no matter how complex the situation – in terms of the number of participants – an individual participant will choose so as to maximize expected utility. In this manner the postulates of the theory of individual, independent decision-making have been regarded as regulative for the theory of interdependent decision-making, at least in the sense of being logically prior to any other behavioural assumptions which might be introduced. Now it follows trivially from the paradigm of individual, independent choice (where the agent is faced with playing a 'game' alone against an 'indifferent' nature) that such a theory can be articulated in abstraction from consideration of possible interdependencies between different agents with respect to their (i) utilities, (ii) expectations, and (iii) decisions. Yet the essence of a game, in the proper sense of that term, would seem to consist precisely in the presence of a complex set of just such interdependencies. The question naturally arises, then: can the behavioural axioms of the theory of individual, independent choice be carried over to, and serve as the foundations for, a theory of interdependent choice?

Attempts to construct the theory of interdependent choice on such a basis have not been marked with unqualified success. Of particular note here are the following considerations:

(i) the recent work of Howard and Schick suggests that there are important solution concepts for a theory of socio-political decision-making which are obtainable only by taking explicit account, in theory construction, of the possible interdependence relations between the utilities of the various participants;

(ii) the lack of consensus with regard to the proper analysis of, and solution to, the 'Prisoner's Dilemma' type of game, and Newcomb's Paradox – the latter constituting an interesting borderline case between independent and interdependent choice – suggests that as yet insufficient attention has been paid to the precise manner in which the expectations and decisions of different participants are related;

(iii) the recent work of Howard, criticizing the 'Sure Thing' Principle, and the work of McClennen, which purports to show that it is impossible to construct an adequate Theory of Games if it is assumed that each player always chooses in accordance with the expected utility hypothesis, suggest that theories which are based on expected utility reasoning are inherently limited or inadequate in certain important respects;

(iv) the ground-breaking work of such persons as Olson, Schelling and Rapoport, on decision-making within the context of social, political and economic conflict, has yet to be fully generalized and made an integral part of a general theory of choice under conditions of interdependency.

Recent treatments of the above-mentioned paradoxes and limitations indicate that there are at least two quite distinct and potentially conflicting remedial approaches. One consists in a reorientation of the Theory of Games away from the postulates of individual, independent decision-making – away from what is essentially a 'reactive' theory in which each participant is to be understood as merely adjusting his choice to his expectations concerning how the others will choose – towards postulates which make the concepts of co-operation and coordinated choice more central. The other approach seeks to resolve the problems mentioned by developing more sophisticated conceptualizations of the postulates of the theory of individual, independent choice as they are applicable to problems of interdependent choice. The aim of this section will be to bring these alternatives into sharp relief and join the issue as to their respective merits.

2.3 *Decision Theory and Normative Theories of Social Ethics and Politics*

Political theory and ethical theory have had a long history of interrelationship. This is nowhere more evident than in the various versions of Utilitarianism, in which the leading conception is that of the organization of social and political institutions with a view to the maximization of the interests or well-being of the persons involved. Such an approach lends itself naturally to a decision-theoretic interpretation, i.e., to interpretation as an application of a general theory of decisions and games. Historically, this approach has stimulated some of the most profound work in the foundations of political and social philosophy, as, e.g., in the works of Hobbes, Hume, Bentham, Mill, Sidgwick, and more recently, Dewey and Rawls.

Economic theory has also been intimately related to political theory, and in this century the foundations for a formal 'economic' justification of political institutions can be traced in the works of, e.g., Baumol, Buchanan and Tullock. The central notion here is that of a resolution of a situation through the creation of a collective social institution that will be capable of enforcing a fair solution. Such institutional coercion is said to be rationally justified because it either removes an external diseconomy or creates or sustains an external economy. The rationality of adopting such an approach, and the precise notion of fairness to which appeal is to be made, has more recently been the subject of considerable debate, as, e.g., in the works of Rawls, Barry, Braybrooke and Gauthier. It should also be noted that the foundational questions to be dealt with in Section 1 have a direct bearing on this issue, since solution concepts appropriate to such situations may be radically different from those permitted by the classical theory, once interdependence conditions are fully integrated into the analysis. It is to be expected, moreover, that such alternative solutions concepts will bear directly on both political and social philosophy in the broad sense (e.g., the possibility of a truly non-individualistic culture) and explanatory models of actual political and social behaviour in our own culture (cf. Section 4).

Economists themselves have always been interested in the theory of decision-making under conflict for their own obvious purposes, but their work also carries across directly to social decision-making in general, in particular to the analysis of collective decisions for a group of persons with disparate individual utilities. This approach has been pursued notably by Arrow, Black and Sen among others. Again, the theoretical work of Section 1 has a direct bearing on the class of problems raised here and is an active centre of investigation at the present time.

The several lines of analysis mentioned above can be seen to converge in a concept which, from the early eighteenth century onwards, has deeply influenced and increasingly dominated Western political and social thought, forming, as it were, the core notion for the conceptualization and justification of the capitalistic industrial state: the idea of a social contract. The focus of this section of the colloquium will be on the development, in recent years, of two radically different and rival versions of such a contractualist theory of ethical and political norms and institutions. The first conceives the social contract decision to be a complex version of a problem in strategic bargaining, in which concepts such as the status quo, threat advantage and strategy play a key role. This approach was historically developed by Hobbes and Hume. The second 'idealizes' the contract decision by conceiving of persons as having to reach a decision from an initial position in which considerations of strategy and relative bargaining power have no place. Its historical roots lie in the writing of Kant and Rousseau but perhaps its clearest formulation is to be found in the contemporary work of Rawls.

Both constructs have been shown to be subject to formidable difficulties (see Arrow, Barry, Braybrooke, Ellsworth, Gauthier, Sen, Sobel).

The bargaining approach turns squarely on the kind of theoretical presuppositions which will be the subject of scrutiny in Section 1 (on theoretical foundations). Moreover, in a series of papers presented at the Canadian Philosophical Association meetings in Toronto in 1974, Gauthier, Sobel and Braybrooke have ably and devastatingly argued that such an approach, if predicated on the presupposition that rational choice is merely expected-utility maximization, leads to distressingly negative results in terms of anything like a traditional contract theory.

The non-strategic, non-bargaining approach manages to avoid these problems. But in so far as it is framed in decision-theoretic terms it must conceptualize the decision on a social contract as a special case of decision-making under conditions of (complete) uncertainty. Once again, serious problems arise – both concerning foundations and with respect to application. The theory of choice under conditions of (complete) uncertainty remains controversial and problematic. (Moreover, recent criticisms of Rawls' *Theory of Justice* (see references above) strongly suggest that the application of concepts of choice under conditions of uncertainty to the social contract problem is beset with considerably more difficulties than were heretofore supposed.)

The aim of this section will be to bring both of the above mentioned approaches into sharp focus and to critically assess their respective strengths and weaknesses.

2.4. *Epistemology and Decision Theory*

The Theory of Rational Belief concerns the epistemological justification of the adoption of particular beliefs and systems of belief. Its historical roots lie in the general problems of epistemology and in the problem of induction (the acceptance of scientific theory on the basis of experimental evidence) in particular. Its more recent development arises from applications of the probability calculus and developments in formal logics for epistemic purposes – founding work in the field having been carried out by Carnap, de Finetti, Keynes, Ramsey, Savage. Closely related to this work is the theory of statistical inference developed in different ways by Fisher, Neyman and Pearson and the followers of the Bayes approach. Specifically, the Theory of Statistical Inference and its relations to Probability Theory have already been made the subject of one of our international colloquia (1976- three volumes, in the University of Western Ontario Series in the Philosophy of Science) and we do not propose to enter upon it here.

Rather, the relevant aspect of this subject is the development initiated by Ramsey and continuing through to Jeffrey and the later Carnap, concerning the relationship between subjective probabilities and epistemic utilities, the development of a coherent set of decision-theoretic rules of epistemic acceptance and the integration of both of these considerations into a Theory of Rational Acceptance of the formal decision-theoretic sort.

In particular, while developing a rich variant of this approach of his own, Levi maintains that epistemic utilities are to be distinguished from practical utilities and treated in separate (sub) systems. This in contradistinction to the Ramsey-Jeffrey tradition. This important epistemic issue will hopefully be confronted directly in seminars involving both Levi and Jeffrey with others who have been actively pursuing these developments – Braithwaite, Churchman, Hempel, Leach, Rudner.

Another important issue to be explored in some detail is the relationship between a Theory of Rational Acceptance as sketched above and a rational reconstruction of the history of science. The tradition that has developed over the past century has led to a cleavage between the abstract, general Theory of Justification for scientific acceptance and change of scientific theory on the one side and sociology and the psychology of scientists as a group of human beings working in particular human cultures on the other side. Corresponding to this cleavage is a radical division of opinion concerning the nature of scientific methodology and in what kinds of terms it is to be characterized. Falling on the traditional side are Hall, and especially the school of thought founded by Karl Popper (e.g., Lakatos, Miller and Watkins). The cleavage in approach arises directly from the characterization of the former theory as an abstract non-individual calculus whose only significant parameters are truth, content and weight of evidence. The tendency in criticism has been to hold the history of science to be beyond rational reconstruction.

The introduction of a Theory of Rational Acceptance, by contrast, at once provides a theory capable of dealing with the actions of individual scientists as rational agents and at the same time introduces a wide range of utilities to the scientific decision-making situation which includes culturally dependent utilities as well as epistemic utilities. Thus the Theory of Rational Acceptance shows promise as being the appropriate basis upon which to develop a unified theory of the development of science (and, incidentally, of scientific institutions – cf. Section 4 below). The colloquium will aim to explore this possibility through confrontation between those writing on the history and methodology of science, on both sides, and those working in Decision Theory.

2.5. *Applications of Decision Theory to the Social Sciences*

As was remarked in the preamble, in recent times there have been developments in all of the social sciences which, while not always consciously recognized as such, have introduced a basic rational model of human behaviour which is in fact most appropriate characterized in decision-theoretic terms. Our aim here is to explore this recent development explicitly.

If we consider traditional Action Theory as it has been developed by the philosophical community, represented, e.g., by Goldman, we find no explicit notion of a theory of rational behaviour at all. And when we turn to the Social Sciences themselves much the same pattern prevails. In the 1950s an important colloquium on human action was convened under the leadership of Talcott Parsons and Edward Shils which brought together representatives of such diverse social sciences as psychology, sociology and anthropology. But the notion of action was there dominated by Parsons' functionalist explanation of action and a theory of rational behaviour was lacking. It is perhaps worth noting in this connection that neither economists nor philosophers concerned with the theory of rational action were present. This particular approach to a unified explanation of human action did not develop fruitfully.

We propose to explore an alternative unified approach to human action, by setting it in a rational decision theoretic context and taking advantage of a number of recent developments.

(i) There is in the Theory of Micro-Economics already an elaborate decision-theoretic account of rational behaviour in certain contexts. Recently this field has been significantly expanded through the work of such persons as Nash, Arrow, Harsanyi, Luce and Raiffa, and Simon.

(ii) Some of the most exciting developments here occurred in the recent expansion of the economic approach to include the functioning of the political structure. We have already seen in Section 2 the recent flourishing of the decision-theoretic approach in respect of questions of social philosophy (justice) and the general 'economic' justification of the state. Recently this work has been extended to include a theory of political decision-making behaviour, e.g., by Buchanan and Tulloch, Downs, Olson, Riker. This theory has in turn raised the problem of interdependent utilities, dealt with in Section 1, in an acute form.

(iii) A further natural extension of the decision theoretic approach is then to the internal structure and functioning of social institutions in general both as organizations and in respect of their individual human member interactions. Such an account, given often in different language, has been recently developed by Blau, Homans, Kuhn (A.), Shubick, Simon and Watkins.

In particular, the work of Homans and Blau returns us also to Section 2 and to the normative/explanatory role for a decision-theoretic model of rational behaviour, for they introduce the principle of distributive justice as an important component of the characterisation of actual expectations and hence as an explanation for actual human behaviour. This use of a rational model perhaps offers the possibility of a more humanist approach to human behaviour than the Skinnerian model (see below) and leads to a new perspective on the conflict between humanism and the social engineering approach. (Perhaps, after all, the two cultures noticed by C. P. Snow is based upon a deep confusion.)

A natural extension of (i) through (iii) is to the Policy Sciences, to the theory of the formation and evaluation of social policy in all areas. As yet this theory does not exist; by and large a theory in this area does not seem sufficiently well articulated, even in its own terms, for one to see clearly the natural way to extend the decision theoretic approach to it. Isolated works that show great initial promise could be cited – e.g., Cole *et al.*, Galbraith, Kuenzlen, Olsson and Schelling – but there is nothing known to us that would justify an additional conference section – rather one might well think in terms of a separate future research colloquium instead.

(iv) Finally there has been a rich anthropological literature developing concerning the issue of inter-cultural comparisons of behaviour in general and the specific understanding of the relations between individual behaviour and social goals, viewed from within a specific culture in particular. Decision Theory shows promise of providing a tool of intercultural understanding – indeed its basic notions have already entered the debate, e.g., in the work of Jarvie and found in the works edited by Buchler, Nutini and Wilson.

(v) Each of these developments makes it increasingly urgent that this general approach to the explanation of human behaviour be confronted with the most important contemporary alternative general approach, that of Skinner's Theory of Operant Conditioning. This theory also promises to give an explanation of the various aspects of human culture, as well as of individual behaviour, and to explain the origin and evolution of values as well (something which a static Decision Theory does not do). In the confrontation between the two one can hope for a fruitful exchange of insights and a better perspective on both. Concerned with this issue have been, e.g., Homans and Simon. In fact, the people working with exchange theory (see Emerson, Homans, Blau) have already brought out something of the relationship of their approach to Skinnerian theory; it remains therefore only to complete the link between exchange theory and decision theory (as in (iii) above), to bring the Skinnerian and decision theoretic approaches to confrontation.

The major aim of the conference in Section 4, then, is to articulate the decision-theoretic approach as a general explanation of human behaviour and to develop it as far as possible, linking it to its foundations, by bringing together contemporary leaders in the field, and to confront the theory with its chief potential rival.

London, Ontario
February, 1977

BIBLIOGRAPHY

Arrow, K., 'Alternative Approaches to the Theory of Choice in Risk-Taking Situations', *Econometrica* **19** (1951).

Arrow, K., *Social Choice and Individual Values*, Yale University Press, New Haven, 1963a.

Arrow, K., 'Uncertainty and the Welfare Economics of Medical Care', *American Economic Review* **53** (1963b).

Arrow, K., 'Control in Large Organizations', *Management Science* **10** (1963–4).

Arrow, K., 'Exposition of the Theory of Choice Under Uncertainty', *Synthese* **16** (1966).

Arrow, K., 'Social Responsibility and Economic Efficiency', *Public Policy* (1973a).

Arrow, K., 'Some Ordinalist-Utilitarian Notes on Rawl's Theory of Justice', *The Journal of Philosophy* 70 (1973b).

Arrow, K., 'Political and Economic Evaluation of Social Effects and Externalities', in Intrilligator M.

Arrow, K., Karlan, S., and Suppes, P., *Mathematical Methods in the Social Sciences*, Stanford University Press, Stanford, 1959.

Arrow, K. and Scitovsky, T., *Readings in Welfare Economics*, Irwin-Dorsey, Homewood, Illinois, 1969.

Bar-Hillel, Y., *Logic, Methodology and the Philosophy of Science; Proceedings of the 1964 International Congress*, North Holland, Amsterdam, 1965.

Barry, B., *Political Argument*, Humanities Press, Boston, 1965.

Barry, B., *Sociologists, Economists and Democracy*, Macmillan, London, 1970.

Barry, B., *The Liberal Theory of Justice*, Oxford University Press, Oxford, 1973.

Blau, P., *Exchange and Power in Social Life*, Wiley, New York, 1964.

Braithwaite, R., *Scientific Explanation*, Harper and Row, New York, 1953.

Braithwaite, R., *Theory of Games as a Tool for the Moral Philosopher*, Cambridge University Press, Cambridge, 1955.

Braybrooke, D., 'Utilitarianism with a Difference', *Canadian Journal of Philosophy* **3** (1973).

Braybrooke, D., 'Will Maximization Lead to the Contract of Self-Defeat?', *Canadian Philosophical Association Institute on Political Philosophy*, 1974a.

Braybrooke, D., 'Another Role for the Contract Idea?', *Canadian Philosophical Association Institute on Political Philosophy*, 1974b.

Buchanan, J. and Tullock, G., *The Calculus of Consent*, University of Michigan Press, Ann Arbor, 1962.

Buchler, I. and Nutini, H. (eds.), *Game Theory in the Behavioural Sciences*, University of Pittsburgh Press, Pittsburgh, 1969.

Buck, R. C. and Cohen, R. S. (eds.), *PSA 1970: Boston Studies in the Philosophy of Science*, Vol. VIII, Reidel, Dordrecht, 1971.

Burger, J., Zelditch, M., and Anderson, D., *Sociological Theories in Progress*, Vol. 2, Houghton Mifflin, New York, 1972.

Burger, R. and Cioffi, F. (eds.), *Explanation in the Behavioural Sciences*, Cambridge University Press, Cambridge, 1970.

Carnap, R., *The Logical Foundations of Probability*, 2nd edition, University of Chicago Press, Chicago, 1962.

Churchman, C. W., 'Statistics, Pragmatics, Induction', *Philosophy of Science* **15** (1948).

Cole, H. S. D., Freeman, C., Jahodo, M., and Pavitt, K. L. R., *Thinking About the Future: A Critique of the* Limits to Growth, Chatto and Windus Ltd., London (for Sussex University Press), 1973.

Colodny, R. (ed.), *Pittsburgh Studies in the Philosophy of Science*, Vol. II, Prentice-Hall, New Jersey, 1965.

Colodny, R. (ed.), *Mind and Cosmos: Explorations in the Philosophy of Science, Pittsburgh Studies in Philosophy of Science*, Vol. III, University of Pittsburgh Press, Pittsburgh, 1966.

Colodny, R. (ed.), *Pittsburgh Studies in the Philosophy of Science*, Vol. IV, University of Pittsburgh Press, Pittsburgh, 1969.

de Finetti, B., 'Foresight: Its Logical Laws, Its Subjective Sources', in Kyburg/Smokler 1964.

Downs, G., *An Economic Theory of Democracy*, Harper and Row, New York, 1957.

Ellsberg, D., 'Risk, Ambiguity, and the Savage Axioms', *Quarterly Journal of Economics* **73** (1961).

Ellsworth, L., 'Rawls' Social Contract Argument', *Canadian Philosophical Association Meetings*, 1973.

Emerson, R. M., 'Exchange Theory', in Colodny, R., 1966.

Feyerabend, P. K., 'Problems of Empiricism', in Colodny, R., 1965.

Feyerabend, P.K., *Against Method* Humanities Press, New Jersey, 1976.

Feyerabend, 'Problems of Empiricism II', in Colodny, R., 1969.

Fishburn, P. C., 'Intransitive Indifference in Preference Theory: A Survey', *Operations Research* **18** (1970a).

Fishburn, P. C., *Utility Theory for Decision Making*, Wiley, New York, 1970b.

Fishburn, P. C., 'Utility Theory With Inexact Preferences and Degrees of Preference', in Leach, J., *et al.*, 1973.

Fishburn, P. C., *Social Choice Theory*.

Galbraith, J. K., *The New Industrial State*, Houghton Mifflin, Boston, 1967.

Galbraith, J. K., *Economics and the Public Purpose*, Houghton Mifflin, Boston, 1973.

Gauthier, D., *Practical Reasoning*, Oxford University Press, Oxford, 1963.

Gauthier, D., 'Morality and Advantage', *The Philosophical Review* **76** (1967).

Gauthier, D., 'Rational Cooperation', *Nous* **8** (1974a).

Gauthier, D., 'The Contract as Ideology', *Canadian Philosophical Association Institute on Political Philosophy*, 1974b.

Hall, R. J., 'Kuhn and the Coperican Revolution', *The British for the Philosophy of Science* **21** (1970).

Hall, R. J., 'Can We Use the History of Science to Decide Between Competing Methodologies' in Buck, R. C. and Cohen, R. S. (eds.), 1971.

Harsanyi, J. C., 'Ethics in Terms of Hypothetical Imperatives', *Mind* **47** (1958).

Harsanyi, J. C., 'On the Rationality Postulates Underlying the Theory of Cooperative Games', *The Journal of Conflict Resolution* **5** (1961).

Harsanyi, J. C., 'Cardinal Utility in Welfare Economics and in The Theory of Risk Taking', *Journal of Political Economy* **61** (1963).

Harsanyi, J. C., 'Bargaining in Conflict Situations in the Light of a New Approach to Game Theory', *American Economic Review* **55** (1965a).

Harsanyi, J. C., 'Cardinal Welfare, Individualistic Ethics and Inter-Personal Comparisons of Utility', *Journal of Political Economy* **63** (1965b).

Harsanyi, J. C., 'General Theory of Rational Behavior in Game Situations', *Econometrica* **34** (1966a).

Harsanyi, J. C., 'A Bargaining Model for Social Status in Informal Groups and Formal Organizations', *Behavioral Science* **11** (1966b).

Harsanyi, J. C., 'Individualistic and Functionalistic Explanations in the Light of Game Theory', in Lakatos, I. and Musgrave, A. (eds.), 1968.

Herzberger, H., 'Ordinal Preference and Rational Choice', *Econometrica.*

Hintikka, J., 'Varieties of Information in Scientific Explanation', in van Rootselaar, B. and Staal, J. (eds.), 1968.

Hintikka, J., 'On Semantic Information', in Yourgrau, W. and Breck, A. D. (eds.), 1970.

Hintikka, J. and Hilpinnen, R., 'Knowledge, Acceptance and Inductive Logic', in Hintikka, J. and Suppes, P. (eds.), 1966.

Hintikka, J. and Suppes, P. (eds.), *Aspects of Inductive Logic*, North Holland, Amsterdam, 1966.

Homans, G. E. C., *Social Behaviour*, Harcourt, Brace and World, New York, 1961.

Hooker, C. A., 'Has the Scientist Any Future in the Brave New World: Copernicus and Velikovsky', *International Symposium Marking the 500th Anniversary of Nicholas Copernicus*, 1973.

Hooker, C. A., 'Systematic Realism', *Synthese* **26**, 1974a.

Hooker, C. A., 'Physics and Metaphysics: A Prologema for the Riddles of Quantum Theory', in Hooker (ed.), *Contemporary Research in the Foundations and Philosophy of Quantum Theory*, Reidel, Dordrecht, 1974b.

Hooker, C. A., 'Cultural Form, Social Institution, Physical System: Remarks Toward a Systematic Theory of Environmental Problems', *Second International Banff Conference Man and His Environment*, Plenum Publishing Company, New York, 1974c.

Hooker, C. A., 'Philosophy and Meta-Philosophy of Science: Empiricism, Popperianism and Realism' *Synthese* **32**, 1975.

Howard, N., *Paradoxes of Rationality: Theory of Metagames and Political Behaviour*, M.I.T. Press, Cambridge, Massachusetts, 1971.

Intrilligator, M. (ed.), *Frontiers of Quantitative Economics*, North Holland, Amsterdam.

Jarvie, I., *The Revolution in Anthropology*, Routledge and Kegan Paul, London, 1964.

Jeffrey, R. C., 'Ethics and the Logic of Decision', *Journal of Philosophy* **62** (1965a).

Jeffrey, R. C., 'New Foundations for Bayesian Decision Theory', in

Jeffrey, R. C., *The Logic of Decision*, New York, 1965b.

Jeffrey, R. C., 'Preferences Among Preferences', *Journal of Philosophy, LXXI*, 1974.

Kuenzlen, N., *Planning Urban Games*, i Press, Boston, 1972.

Kuhn, A., *The Study of Society*, Irwin Dorsey, Homewood, Illinois, 1963.

Kuhn, T., *The Structure of Scientific Revolutions*, University of Chicago Press, Chicago, 1962.

Kyburg, H. E. and Smokler, H. E., *Studies in Subjective Probability*, Wiley, New York, 1964.

Lakatos, I. and Musgrave, A. (eds.), *Problems in the Philosophy of Science*, North Holland, Amsterdam, 1968.

Lakatos, I. and Musgrave, A. (eds.), *Criticism and the Growth of Knowledge*, Cambridge University Press, Cambridge, 1970.

Leach, J., 'Explanation and Value Neutrality', *British Journal of the Philosophy of Science* **19** (1968).

Leach, J., Butts, R., and Pearce, G. (eds.), *Science, Decision and Value*, Reidel, Dordrecht, 1973.

Lehrer, K., 'Induction: A Consistent Gamble', *Nous* **3** (1969).

Lehrer, K., 'Justification, Explanation and Induction', in Swain, M. (ed.), 1970a.

Lehrer, K., 'Induction, Reason and Consistency', *British Journal for the Philosophy of Science* **21** (1970b).

Levi, I., 'Must The Scientist Make Value Judgements?', *Journal of Philosophy* **67** (1961).

Levi, I., 'On the Seriousness of Mistakes', *Philosophy of Science* **29** (1962).

Levi, I., *Gambling with Truth*, A. A. Knopf, New York, 1967.

Luce, R. D. and Raiffa, H., *Games and Decision: Introduction and Critical Survey*, Wiley, New York, 1967.

McClennen, E., 'An Incompleteness Problem in Harsanyi's General Theory of Games and Certain Related Theories of Non-Cooperative Games', *Theory and Decision* **2** (1972) 314–341.

Miller, D. W., 'On the Comparison of False Theories by Their Bases', Warwick, England, mimeographed.

Miller, D. W., 'Popper's Qualitative Theory of Verisimilitude', Warwick, England, mimeographed.

Nash, J. F., 'The Bargaining Problem', *Econometrica* **18** (1950).

Nash, J. F., 'Non-Cooperative Games', *Anals of Mathematics* **54** (1951).

Nash, J. F., 'Two Person Cooperative Games', *Econometrica* **21** (1953).

Nozick, R., 'Newcomb's Problem and Two Principles of Choice', in Rescher, N. (ed.), 1970.

Olsen, M., *The Logic of Collective Action*, Schocken, New York, 1971.

Olsson, G., 'Correspondence Rules and Social Engineering', in *Economic Geography* **47** (1971).

Olsson, G. (ed.), 'Geography, Epistemology, and Social Engineering', *Antipode* **4** (1972), special issue.

Parsons, T. and Shils, E. A. (eds.), *Toward A General Theory of Action*, Harper and Row, New York, 1951.

Popper, K. R., *The Logic of Scientific Discovery*, Hutchinson, London, 1962.

Popper, K. R., *Objective Knowledge*.

Radner, M. and Winokur, S. (eds.), *Minnesota Studies in the Philosophy of Science*, Vol. IV, University of Minnesota Press, Minneapolis, 1970.

Ramsey, F. P., *The Foundations of Mathematics and Other Logical Essays*, Routledge, London, 1931.

Ramsey, F. P., 'Truth and Probability', in Ramsey, F. P., 1931, and Kyburg, H.E. and Smokter, H. E., 1964.

Rapoport, A., *Two-Person Game Theory: The Essential Ideas*, University of Michigan Press, Ann Arbor, Michigan.

Rapoport, A., *N-Person Game Theory: Concepts and Applications*, University of Michigan Press, Ann Arbor, Michigan.

Rapoport, A., *A Strategy and Conscience*, Harper and Row, New York, 1964.

Rawls, J., *A Theory of Justice*, Harvard University Press, Boston, 1971.

Rescher, N. (ed.), *Essays in Honour of Carl G. Hempel*, Reidel, Dordrecht, 1970.

Riker, W. H., *The Theory of Political Coalitions*, Yale University Press, New Haven, 1962.

Riker, W. H. and Ordeshook, P. T., *An Introduction to Positive Political Theory*, Prentice-Hall, New York, 1973.

Rudner, R., 'The Scientist Qua Scientist Makes Value Judgements', *Philosophy of Science* **20** (1953).

Savage, L. J., *The Foundations of Statistics*, Wiley, New York, 1962.

Schelling, T. C., *The Strategy of Conflict*, Harvard University Press, Boston, 1960.

Schelling, T. C., 'On the Ecology of Micro-Motives', *The Public Interest*, **25** (1961).

Schick, F., 'Arrow's Proof and the Logic of Preference', *Philosophy of Science* **36** (1969).

Schick, F., 'Beyond Utilitarianism', *The Journal of Philosophy* **68** (1971).

Schick, F., 'Democracy and Interdependent Preferences', *Theory and Decision* **3** (1972).

Sen, A. K., *Collective Choice and Social Welfare*, Holden-Day, San Francisco, 1970.

Shapere, D., 'The Structure of Scientific Revolutions', *Philosophical Review* **73** (1964).

Shapere, D., 'Meaning and Scientific Change', in Colodny, R. (ed.), 1966.

Shubick, M. (ed.), *Game Theory and Related Approaches to Social Behaviour*, Wiley, New York, 1964.

Simon, H. A., *Models of Man*, Wiley, New York, 1957.

Simon, H. A. and March, G., *Organizations*, Wiley, N.Y., 1968.

Skinner, B. F., *Science and Human Behaviour*, MacMillan, New York, 1953.

Sobel J. H., 'The Need for Coercion' *Nomos* **14**, 1972.

Sobel, J. H., 'Interaction Problems of Utility Maximizers: Their Nature, Avoidance and Significance', *Canadian Philosophical Association Institute on Political Philosophy*, 1974a.

Sobel, J. H., 'Defeat of Utility Maximization in Iterated Prisoner's Dilemma: The Fact and Its Significance', *Canadian Philosophical Association Institute on Political Philosophy*, 1974b.

Swain, M. (ed.), *Induction, Acceptance and Rational Belief*, Reidel, Dordrecht, 1970a.

van Rootselaar, B. and Staal, J. (eds.), *Logical Methodology and Philosophy of Science*, Proceedings of the 3rd International Congress, North Holland, Amsterdam, 1968.

von Neumann, J. and Morgenstern, O., *Theory of Games and Economic Behaviour*, 2nd edition, Princeton University Press, Princeton, New Jersey, 1947.

Watkins, J. W. N., 'Decisions and Uncertainty', *British for the Philosophy of Science* **6** (1955).

Watkins, J. W. N., 'Imperfect Rationality', in Burger, R. and Cioffi, F. (eds.), 1970.

Wilson, B., *Rationality*, Harper and Row, New York, 1970.
Yourgrau, W. and Breck, A. D. (eds.), *Physics, Logic and History*, Plenum Press, New York, 1970.

ACKNOWLEDGEMENTS

The Editors would like to thank the Faculties of Arts and Social Science of The University of Western Ontario, and also the Canada Council, for financial support of the workshop. We are also much indebted to Alice Smith for an excellent job of preparing and typing the collected manuscripts.

Our gratitude, too, to Andrew Bjerring for his work in preparing the index.

DAVID BRAYBROOKE

POLICY-FORMATION WITH ISSUE-PROCESSING AND TRANSFORMATION OF ISSUES

Adopting a certain model for policy-formation, one may impute to a stable region of the political system an issue-machine, i.e., the analogue of a computer, with a program in place for processing issues. The 'computer' will consist of a network of agencies or stations, public or private. The program will call upon them (perhaps repeatedly) in a certain sequence and (giving effect to their reactions) round by round process issues of a given sort in much the same way, with much the same results.[1]

One way of conceiving of the issues going through the machine is to conceive of them as having the form of a certain type of question – in Belnap's treatment of the logic of questions, the form of a multiple-answer whether question offering a disjunction of policy proposals P_1 or P_2 or . . . P_n.[2] Conceived in this way, an issue may be said to be of one sort rather than another because the disjunction of policy proposals (each indifferently expressed by any of a class of synonymous locutions) that it offers consists wholly of logically suitable answers to a distinctive issue-circumscribing question.[3] This question in its turn may be thought of as having a special form in the logic of questions – most conveniently, it seems to me, the form (to use Belnap's terminology again) of a multiple-answer which question.[4]

Such a question does not itself formulate alternative possible answers, but stipulates a condition or conditions that answers must meet. For example, the question, 'What shall we do to multiply parking places in the central city?', circumscribes issues of a different sort from those circumscribed by the question, 'How can we check the growth of traffic congestion in the central city?' To answer the first question, a proposal must meet (or at least seem to meet) the condition of making more parking places available. To answer the second question, a proposal must meet or seem to meet the condition of checking the growth of traffic congestion. Logically, a proposal might meet one and not meet the other. (Causally, too, as it happens, proposals answering the first question fail as answers to the second question.)

During the passage of an issue through the issue-machine, some, perhaps all, of the proposals in the issue as originally posed may be eliminated. Bearing upon them in due order and due time, the various contravening

Hooker, Leach, and McClennen (eds.), Foundations and Applications of Decision Theory, Vol. II, 1–16. *All Rights Reserved.*

reasons invoked in the program cumulate in rejecting them.

If all of them are eliminated, and no further proposals are added, the issue-machine stops short of producing an outcome. The issue becomes a dead one. From another point of view, of course, we may choose to say that the outcome is the *status quo*. However, in fact no attention may have been given to the *status quo* as one of the disjuncts (proposals) in the issue. This inattention may be significant – there may, for example, have been no partisans of the *status quo*.

Often there is an outcome. One of the proposals in the issue runs the gamut of reasons successfully and becomes a policy of the political system to which the issue-machine belongs. Maybe more than one does. It accords best with issues actually observed to refrain from assuming that the alternatives discussed are mutually exclusive.

Moreover, we may have in view an issue-machine that provides for the introduction of further proposals during the period of processing. Some of these may win through to become outcomes, even if none of the original proposals do.

Every addition of a distinct proposal to the disjunction being processed may be looked upon as a transformation of the issue. The issue may continue to be an issue of the same sort; but it becomes a different issue nevertheless.

Every subtraction may be looked upon as a transformation, too. Transformations by subtraction, however, are automatically provided by the concept of processing. I shall put them aside during the following discussion.

I shall also put aside for most of the way on my present trip, the transformations, potentially more radical and inherently most interesting of all, produced by changing the issue-circumscribing question. Consider, for example, how an issue about the use of parking meters might be transformed by changing the issue-circumscribing question, as above, from 'What shall we do to multiply parking places in the central city?' to 'How can we check the growth of traffic congestion in the central city?'

I am to a degree overdramatizing the historical example that I have in mind. What actually happened in England was that proposals answering the first question were first expected to help answering the second question, or at least to be compatible with answering it; later they were seen to be incompatible. The second question was always in the field, jointly important if not supremely so. Yet after it was recognized that to multiply parking places aggravated congestion, the first question vanished and with it the particular issues that it had generated and circumscribed.

Sometimes the new light in which a given disjunction of policy-proposals appears, once the issue-circumscribing question has been changed, will lead, through changes in people's preferences for those proposals, to a decision, without rendering any of the given proposals irrelevant and without generating any new proposals for consideration. I suppose, however, that changing the issue-circumscribing question will usually change the issue as well, producing among its effects simpler transformations, by subtraction and addition. (I take completely replacing the proposals in a given issue by an entirely different set to be a limiting case simultaneously of both subtraction and addition.) Moreover, it is only by way of these transformations that changing the issue-circumscribing question will make any difference to deciding the issue, if people's preferences for the proposals first given remain the same. To keep the present inquiry within manageable limits, I shall assume, most of the way, that they do remain the same. With that assumption, the simpler transformations are in a sense basic, as well as being on several grounds a natural place to begin. Among these transformations, I shall, as I said, concentrate on transformations by addition.

Transformations by addition are more interesting for my present purposes because they present an opportunity to discuss a neglected technique for resolving issues. It is a technique that the formal theory of collective choice continually neglects. Yet it is a technique so much used, and so prominent, as (I think) to overshadow in practical importance the techniques which the formal theory of collective choice does treat.

(To discuss this technique, it is not strictly necessary to postulate an issue-machine. Its role in the transformation of issues, indeed the whole subject of such transformations, in all its variety, can perhaps be investigated at length without the help of the issue-machine idea. However, an issue-machine supplies a congenial context. If one is not postulated, transformation by subtraction becomes hard to explain; and if one is postulated, the same basis for explanation readily extends to the other sorts of transformation. If we ask, which transformations falling under these sorts have in given circumstances which effects on the fate of given issues, we are naturally led to talk about the reasons and actions of various participating agencies and stations.)

Consider the following sorts of transformation by addition: (1) adding to the existing disjuncts – the proposals being entertained – a new proposal or disjunct that combines the chief attraction of P_1, favored by one group, with the chief attraction of P_2, favored by another; (2) adding a proposal with an attraction or attractions that differ from those of P_1 and P_2; (3) adding a proposal that is inferior to P_1 respecting its chief attraction, but in this respect superior to P_2, and inferior to P_2 respecting the chief attraction of P_2, but in this respect superior to P_1. In the last case, we have a typical instance of one sort of compromise.[5]

The current theory of collective choice can say something about all these cases (once the additions in question have taken place). It would treat the first case as one in which (other things being equal) unanimity can be obtained by moving directly to a position superior by the Pareto welfare criterion. It might treat the third case as one in which unanimity may be obtained under the pressure of bargaining. The two parties to the conflict, if they both accept the added proposal, do so because they each decide, given the resistance which they have encountered, that they will not be able to gain more for themselves. Just what the theory might do about the second case is not so clear, but, sometimes at least, this case may be viewed as one in which the added proposal succeeds if it does, because it in some way compensates both parties, one for the loss of P_1, the other for the loss of P_2.[7]

To say no more than this about the three cases, however, is to miss their significance as illustrations of a particular technique for resolving issues. To say no more than this passes over the crucial difference between having a set of proposals given with an issue and bringing another set of proposals into view by transforming the issue.

Arrow (followed by Sen and others) deliberately puts the difference out of sight for theoretical purposes. He expressly chooses to deal, in the theory of collective choice, only with issues in which the alternatives are given from the beginning. Furthermore, he increases the distance at which he operates from the issues and techniques of real politics by insisting also on dealing with issues in which the alternatives involve whole social states, given and completely described.[8]

I am not denying that with these assumptions Arrow obtained brilliant results and inspired illuminating work by others. I would contend that the strong light which he has thrown upon part of the subject of collective choice has been dazzling and distracting to an extent not commonly realized by those working in the light. At any rate, a substantial part of the subject has been cast into relative obscurity.

To describe a social state 'completely', Arrow does not ask that every feature which might be predicated of it be expressed, something that may well be unintelligible, and certainly impossible to accomplish. He means that the quantities of resources allocated to different productive activities should be specified, along with the quantities of goods and services finally distributed to individual persons or given a collective use.

I am not sure that even this sort of description is feasible, without leaving out details which make a difference to individual consumers. However, let us suppose it is. To be prepared, as Arrow is prepared, to treat issues about whole social states, completely described in the alternatives proposed, may seem at the same time to offer sufficient room for treating any issue about variations in the details of social states and to eliminate any need for considering variations in alternatives beyond what is given.

For cannot choosing between variations in detail be treated as equivalent to choosing between social states varying in just those details? Suppose, then, that those variations – and any others that one might wish to specify – are given, with whole social states embodying them.

Yet, even at the elevated position where Arrow chooses to raise problems of collective choice, not every issue about given proposals is necessarily given with them; nor is every proposal that a given issue is capable of embracing. There is room for politicians and others to work for the transformation of issues.

Given any number of whole social states, completely described, that have been put before the public as alternatives, further, non-synonymous descriptions may be offered of them. The social state called for by P_1, for example, may receive, besides its complete economic description, the description of injustice to some social class. The social state called for by P_2 may at the same time receive the description of injustice to another class, or of entailing betrayal of some international obligation. A further proposal, P_3, that wins decisive support may do so because it answers an issue about justice or international obligation (or both), introduced with these descriptions, not because it is Pareto-optimal in relation to P_1 and P_2, expressed in economic terms. Given yet other descriptions of the social states involved in the three proposals, P_3 may win because its attractions, if they are to be regarded as compensations at all, compensate in other than economic terms; or because one or another of the parties in a bargaining situation, seeing how resolute the other party or parties are, is brought to look upon the proposal that it was backing as portending oppression.

These effects arise from transforming the original issues by transforming the issue-circumscribing questions under which the issues first appeared. It is not my present business to discuss them further, since my chief present purpose is to discuss transformations by addition.[9] However, transformations by addition may also be available at the elevated position where only proposals about whole social states appear as the elements of issues. (I am waiving for the moment the assumption that *all* such states appear.) Consider P_1 and

P_2 as examples of the proposals originally given, both involving complete descriptions of whole social states in economic terms. The descriptions will mention the quantity of every good and service distributed for private consumption, $G_1, G_2, \ldots, G_n$, indicating the amount of each received by each consumer, and every collective good, $C_1, C_2, \ldots, C_n$, with the amount produced of each. Whatever the original issue-circumscribing question may be, suppose that it remains unchanged.[10] If someone now invents a new good G_{n+1} or new collective good C_{n+1}, new social states can be proposed, with various amounts of these new goods. (The social states formerly in view will have to be redescribed to keep their title to being 'completely' described; one may simply add to each of the original descriptions the statement that the amount of G_{n+1} and of C_{n+1} supplied is zero.)

Transformations by addition in which further whole social states are brought into review are ruled out of Arrow's theory by the further, heroic assumption, which stretches what is supposed to be given to embrace 'all conceivable social states' and postulates that 'each individual in the community has a definite ordering' of them all 'in terms of their desirability to him'.[11] At any rate, the assumption rules out such transformations if we take 'conceivable' to mean 'conceivable at any time, now or in the future'.

The assumption does not rule out transformations respecting whole social states like those mentioned just previously, in which new descriptions are furnished of states already given with descriptions in the economic terms that Arrow has in mind. At any rate, it does not do so if it is not pushed – further than Arrow gives any sign that he wants to push it – in the direction of hyperfantasy about commanding all conceivable descriptions of all conceivable social states.

The assumption is sufficiently unrealistic as it stands. The extent to which it precludes transformations even of elevated issues about whole social states offers a partial measure of the lack of realism inherent in assuming that all proposals are given.

We cannot at any time hope to specify all the goods that might be thought to figure in a social state, and hence we could not completely describe all social states (or indeed, completely describe any, free of possible revision), unless we used catch-all categories that masked the introduction of novel goods. (Imagine using 'miscellaneous' to mask the introduction of the automobile! 'Private transport' would not be much better.) Moreover, additions of the kind in question are not confined to material technology. (Even if they were, they could not be realistically treated as originating wholly outside politics.) The sphere of collective activities offers politicians considerable opportunities for invention. They may think up new circuses; new ceremonies; new programs for enlisting civic participation; new schemes for insuring the hitherto uninsurable.

To get a full measure of what the assumption that all proposals are given sacrifices in realism, however, one must descend from the elevated position (never reached in actual politics) where whole social states are contemplated among the alternative proposals.

Perhaps Arrow would not wish to keep the assumption, descending. He would nevertheless only be following a familiar precedent if he did. For the problems that the theory of choice, collective and individual, commonly sets itself do not allow for variations in the set of alternatives faced, once the set is given.

Whatever Arrow would decide to do in descending, it is to the point for us to ask just what sorts of activities and techniques for resolving issues he left below during his ascent.

Again I wish to concentrate on transformations by addition. Leaving room for other sorts of transformations to figure in the process, let us call the process by which further proposals intended to supersede in attractions the proposals already present are added to an issue-disjunction, 'The Revisionary Process'. The process is a very basic feature of politics – logically more basic, for instance, than the strategy of disjointed incrementalism, since other features as well enter into the strategy.[12]

We have just seen that there is considerable room for the Revisionary Process to operate even should the proposals in view differentiate whole social states, completely described (so long as we do not suppose, hyperfantastically, that all such descriptions are given). The room for the process expands exponentially when we descend to issues of an ordinary sort, where policy proposals – about parking places, traffic congestion, public transport – do not present anything like completely described whole social states. They present, and they are understood to present, alternative ways of answering issue-circumscribing questions about specific subjects – like parking or traffic congestion – leaving manifold other subjects out of account.

That is not to say that the other subjects are ruled out as irrelevant, much less that they are assumed for purposes of discussion not to vary. Such moves, convenient as they may be for theory, pass over important features of ordinary practice. In particular, they divert attention from the work that politicians and others may do in making connections with other subjects and bringing other subjects up in arguments for or against various proposals. For example, certain proposals for checking the growth of traffic congestion may be argued, quite unexpectedly, to have effects on the distribution of real income among motorists.

It is at least in part by exploring other subjects that politicians and others find additional proposals to put forward in the Revisionary Process, as applied to a given issue about a specific subject. No doubt some additional proposals can, on occasion, be found by formulating variations on the terms of the proposals already given. The third of the three cases of addition originally given – where P_3 excels P_1 in respect of the special attraction of P_2 and excels P_2 in respect to the special attraction of P_1 – may readily be thought of as produced by such a maneuver. The second case of addition, however, in which P_3 has attractions of a different kind from those offered by P_1 and P_2, may well require attention to some other subject. For example, P_3 will not only relieve traffic congestion to some degree, but will tend to make the distribution of real income more equitable. The first case of addition, too, may well bring in other subjects. There P_3 combines the chief attractions of P_1 and P_2; and to combine them one may have to make important innovations elsewhere (and maybe start up other issues there). For example, it may be proposed that road building be funded, under a special roads authority, separately from other operations of government. Then, if the funds are collected by charging motorists more for driving on congested roads, two attractions could be realized – both having the roads used in an optimal way and having more of them built.

Another reason for insisting on keeping in view the fact that ordinary politics deals with specifically circumscribed issues, far short of issues about whole social states, is that it helps keep one in mind of the fact that real-world politics takes up issues seriatim. Real-world politics does not seek comprehensive solutions that will stand indefinitely unless people's preferences should (perhaps quite randomly and irrationally) change. It leaves room for future adjustments to the present subject along with room for future partial settlements of subjects other than the present one.[13] But this is further room for the work of politicians and for the Revisionary Process. P_3, a proposal more satisfactory to most people than any of the proposals raised in the present discussion, may come up later and when it comes supersede the policy adopted now.

The Revisionary Process can be looked upon as a procedure for arriving at choices of social policy. Sometimes, for lack of imagination on the part of the participants or through defects in their institutions, it is not exploited very effectively. Even so, if we take into account the whole range of social choices, governmental and extra-governmental, we may reasonably say that the Revisionary Process often succeeds in arriving at a proposal generally or unanimously agreed to be more attractive than any hitherto offered as an answer to a given issue. Moreover, this proposal often meets the minimum standard of acceptance that is embodied in the program of the relevant issue-machine or is otherwise imposed by the sentiments currently prevailing in the political community. On such occasions there is little left for formal procedures of social choice to do but ratify the proposal as the policy to be followed – at least for the time being.

This way of preparing for social choice and finally making it will (like any other) fail to meet Arrow's conditions for passing from personal preferences to social choice. I think its failure to do so is not after all an important practical embarrassment, as Arrow himself is in a position to admit, though he firmly refuses to carry on his theoretical inquiry at a level where the Revisionary Process and its contribution to resolving issues become visible.[14] Arrow does not deny that it is frequently possible to arrive at satisfactory choices of social policy on issues of limited scope. He holds that no procedure is vindicated by such results in the face of the conditions that he would impose on social choices ranging over all possible configurations of personal preference-orderings.

In view of the scope for the Revisionary Process in real-world politics, however, the absence of practical embarrassment in its application implies that the difficulties in the theory of social choice so startlingly demonstrated by Arrow are more *recherché* than he would care to admit. Tullock, citing the scope for log-rolling within the scope of what I call the Revisionary Process, tried to make this point.[15] He slipped up, unfortunately, in not giving due attention to Arrow's assumptions, which leave no scope for log-rolling. As Arrow was impelled to remind him, social choices with which Arrow is concerned are to take place between whole social states, with 'all possible combinations of decisions on the separate issues . . . considered as alternative social states'.[16]

Tullock's aim was off. Nevertheless the target is there and he had seen it. He could have taken aim again, this time at Arrow's assumptions, with a mind to showing the distance to which they elevate Arrow's conception of social choice above the preoccupations and techniques of real-world politics.

Ex post, all the three cases of transformation by addition that I have used to illustrate the Revisionary Process are subject to Arrow's Impossibility Theorem. The choice emerging after the transformation cannot be a choice by a procedure meeting his conditions, because there is no such procedure. On this point, Arrow's proof applies outside his assumptions about choosing among whole social states, although its effects, properly viewed, are not there so disturbing. *Ex post*, of course, all the proposals that were considered in passing to the social choice have become known. We can say that whatever procedure was used to choose among them violated one or another of Arrow's conditions.

By hypothesis, not all the proposals were known *ex ante*. However, Arrow's Theorem applies *ex ante*, too, in the sense that we know in advance that whatever proposals are brought forward, radically and imaginatively revised though some of them may be, great feats of political invention, the procedure for reaching the social choice among them will not meet Arrow's conditions.

These conditions are in some respects excessively weak. A procedure that met the non-dictatorship condition, for example, might be unattractively oligarchical.[17] The conditions are also, however, excessively strong. They disqualify, for example, any procedure with which even the possibility arises of encountering the paradox of voting. As Tullock again (among others) has pointed out, we might put up comfortably enough with procedures that rarely in fact produce such paradoxes.[18] We had better say, rarely in fact force us to confront such paradoxes, since it can be shown that the chance of there being a winning alternative not involved in a paradox moves within vanishing distance of zero as the number of alternatives increases without limit.[19] In real-world politics, fortunately, there is no incentive to increase alternatives without limit.

In the respects in which Arrow's conditions are excessively strong, the Revisionary Process, when it succeeds as it might succeed in the three standard cases I have mentioned, need not be embarrassed by failures to meet them. When it so succeeds, it absolves any procedure with which it is associated from creating an actual paradox, at least as regards fixing upon an unambiguous first-choice policy.[20] When it succeeds repeatedly, it reduces the practical embarrassment associated with not meeting Arrow's conditions toward the vanishing point. Moreover, the procedure may also (though this would remain to be established) be steadily democratic, in respect to affording a reasonably extended opportunity for every segment of the population to offer proposals and encouraging everyone involved to aim at consensus.

Often, in the last stage of arriving at the resolution of a political issue, some body of people is called upon to vote, using some system of voting; and sometimes, they have more to do than merely ratify the resolution of the issue reached by the Revisionary Process beforehand. Arrow's Theorem and the theoretical developments that it has stimulated have a special application to real-world politics here, besides offering general illumination on what aggregating personal preference orderings might signify.

May not even this special application, however, unless very judiciously managed, lead to misinterpreting the significance of the voting in question? For how often does voting, even when it has a more important part to play than merely ratifying, amount to attempting to establish a final firm first-choice among policies, much less a more extensive ordering of the extant proposals? I would suggest that, more often than not, even such voting rather signifies that in the opinion of some decisive set the time has come to bring the Revisionary Process to an end, for the time being, in application to a certain issue. It is pressingly felt that other issues must now be dealt with. Quite consistently with this opinion, the decisive set might believe that even more attractive proposals might be forthcoming if the Revisionary Process had been continued longer, or were resumed, as sometime in the future it might be expected to be.

Insofar as real-world politics is understood by the participants in such a light, an *ex ante* view of given issues persists even after they have been – for the time being – resolved. Moreover, it is an *ex ante* view that makes the sort of problem which the Revisionary Process copes with so different from the sort of problem posed by issues of the kind that Arrow envisages in his elevated position. Real-world politics sets up issues so as to leave room for new proposals to be found by political innovation, at the very least through transformation by addition. (Again, I am not saying that we can count on our institutions to use the room imaginatively.)

One might say that real-world politics poses problems that can be solved in two ways – by answering them as they stand; or by finding problems that are easier to answer. The second way, the native home of the Revisionary Process, has been lost from sight in the formal theory of collective choice.

Arrow, of course, is not alone in neglecting it. Neglect of it is endemic in the current formal theory of rational choice, personal as well as social. What could be more inviting, as a theoretical simplification, than to assume that if choices are going to take place between specifiable alternatives, they can be regarded as taking place between alternatives already specified? Common use shows how inviting.[21] But that is not the way social choices typically work – or, for that matter, many or most personal choices either.[22]

Dalhousie University

NOTES

[1] Cf. my *Traffic Congestion Goes Through the Issue-Machine*, Routledge, London, 1974.

[2] Nuel D. Belnap, Jr., *An Analysis of Questions: Preliminary Report*, System Development Corporation, Santa Monica, 1963. (Technical Memorandum 1287/000/000.) Now available, essentially unchanged, as part of Nuel D. Belnap, Jr., and Thomas B. Steel, Jr., *The Logic of Questions and Answers*, Yale University Press, New Haven, 1976.

[3] We leave it to the political community to decide (it may be, after discussion) which are synonymous locutions, as not differing in any politically significant way.

[4] Issue-circumscribing questions, with such a form, are also called 'issues' in ordinary language. Instead of distinguishing between issues and issue-circumscribing questions, I might have distinguished between two types of issue, with different forms: generated issues, perhaps, versus generating ones.

[5] 'Compromise' seems to me to be a grossly understudied concept, so far as study of it requires analyzing familiar notions of compromise and relating them to sophisticated discussions of bargaining and social choice.

[6] Cf. Amartya K. Sen, *Collective Choice and Social Welfare,* Holden-Day, San Francisco, 1970, on the first and third cases, p. 26.

[7] If one or the other of the two parties has multiple members, a decision may be reached short of unanimity, in any of these cases, by some variation on majority rule. The theory will explain how the majority subset reaches unanimity on the same lines that it explains unanimity overall, when unanimity occurs.

[8] Kenneth J. Arrow, *Social Choice and Individual Values,* 2nd edn., p. 17, p. 109. Wiley, New York, 1963; cf. Sen, *loc. cit.*

[9] On one view of the individuation of actions, events, and (we may add) states of affairs, the transformations just described could be treated as transformations by addition; for there would be as many social states on this view as there are non-synonymous descriptions. The polar opposite of this view, of course, is the view that any given social state (like any given action or event) is susceptible of infinitely many true non-synonymous descriptions. A good deal of recent philosophical discussion has revolved about positions between these extremes; for a recent contribution, which gives a comprehensive review of the literature, see Monroe C. Beardsley, 'Actions and Events: The Problem of Individuation', *American Philosophical Quarterly,* Vol. XII, No. 4, pp. 263–276, October 1975. I am myself inclined toward the second pole, freely allowing multiple descriptions of the same social state. However, even people inclined toward the first pole, and consequently inclined to find transformations by addition where I would not, should bear in mind that such transformations would in cases like those I have described almost surely lead to transformations of the issue-circumscribing question also. (I am indebted to my colleague Alexander Rosenberg for a useful discussion bearing on this note.)

[10] The original issue-circumscribing question might, compatibly with Arrow's approach, be a question about justice or international obligations or non-oppression. Personal orderings vary, on his account, with moral values as well as tastes; and, in special cases, all the personal orderings might be determined by views of justice or international obligations or non-oppression. Cf. Arrow, *op. cit.*, pp. 17–18. The general case, however, is one in which values vary greatly among persons, so much so that the only common

issue-circumscribing question may be one about the social ordering to adopt, given the personal orderings (i.e., 'Which Bergsonian welfare function should we choose, given these personal orderings?'). On this point, Arrow might claim with some foundation to be realistic. Nevertheless, besides neglecting transformations by addition and the work that politicians do in producing such transformations, Arrow's approach leaves out of account the work that they may do, with or without transformations by addition, in changing issue-circumscribing questions. Partly by introducing new descriptions of proposals, they may seek to bring about convergence upon a specific issue-circumscribing question, as a step toward bringing about decisive support for a proposal answering to an issue so circumscribed.

[11] Arrow, *op. cit.*, p. 17.

[12] For the place of the Revisionary Process in disjointed incrementalism, see the passages in D. Braybrooke and C.E. Lindblom, *A Strategy of Decision,* The Free Press, New York, 1963, on 'reconstructive treatment of data', pp. 98-99, and on supersession and re-designing, pp. 194-197.

[13] That politics is a continuing process of partial adjustments is, of course, one of the chief themes of *A Strategy of Decision.* Lindblom has returned to it in other works, notably *The Intelligence of Democracy,* The Free Press, New York, 1965.

[14] Arrow, *op cit.*, p. 109.

[15] Gordon Tullock, 'Theoretical Forerunners', Appendix 2 in J. M. Buchanan and Gordon Tullock, *The Calculus of Consent,* p. 332, University of Michigan Press, Ann Arbor, 1962 (cited by Arrow, *loc. cit.*). One may be inclined to think of log-rolling as occurring across different issues, but it can also be found taking place within an issue, circumscribed as I am assuming an issue would be. Let F_1 be a feature of some policy, and a feature attractive to one party; let F_2 be another such feature, attractive to another party. Then one party may say, 'We'll accept F_2 if you'll accept F_1', and the other party may agree. P_3 in any of my three standard cases may be the policy that combines F_1 and F_2, depending on further details about the character of those features; and the issue to which it is added may remain circumscribed as it was initially.

[16] Arrow, *loc. cit.*

[17] Cf. Sen, *op. cit.*, pp. 49-50.

[18] Tullock, *op. cit.*, p. 334.

[19] I am relying for this point chiefly on an article, then unpublished, by Charles R. Plott ('Axiomatic Social Choice Theory', 1975), presented at the Key Biscayne conference on Social Choice Theory and Democratic Theory sponsored by the Mathematical Social Sciences Board in December, 1975. But see William H. Riker and Peter C. Ordeshook, *An Introduction to Positive Political Theory,* p. 96, Prentice-Hall, Englewood Cliffs, N.J., 1973, and the references there cited. Plott's article has subsequently appeared in *American Journal of Political Science*, Vol. XX, No. 3, pp. 511–596, August 1976.

[20] Paradoxes as regards *n*-tuples of proposals other than the first choice may appear in the disjunctions added to by the Revisionary Process, even if there were no such paradoxes as regards the original proposals. On the hypothesis that ordinary people and politicians would not be much disturbed by them, or feel impelled to resolve them, disturbed or not, I am not counting them as actually confronted. Politicians may sometimes have opportunities and incentives to force such confrontations, sometimes by adding proposals deliberately intended to undermine the paradox-free claims of an otherwise winning alternative. I am throughout disregarding these possibilities. It clearly exagger-

ates the advantages of the Revisionary Process to assume (in effect) that the people taking part in it are all sincerely trying to find a policy attractive to all sides. For a useful discussion of manipulative tactics in paradox-prone situations, see Riker and Ordeshook, *op. cit.*, pp. 94–100.

[21] For an up-to-the-minute example of standard current thinking at the highest level of competence, see Brian Barry and Douglas W. Rae's article, 'Political Evaluation', in Fred I. Greenstein and Nelson W. Polsby (eds.), *Handbook of Political Science*, Vol. I, pp. 337–401, Addison-Wesley, Reading, Mass., 1975. They do at one place (p. 350) note that it may be possible to 'expand the set of available choices from which an acceptable one may emerge'. However, their general perspective is one in which the difficulties of evaluating and choosing policies, and the opportunities for overcoming those difficulties as well, arise for procedures themselves incapable of expanding the set of alternatives. Thus they argue (p. 348) that it is no use settling for 'a procedure that will always produce a single preferred alternative (or more than one in equal first place) even if it cannot give a consistent ordering of the other possibilities . . . because any procedure capable of always providing a first choice from any list of alternatives must be capable of always providing a complete ranking'. They do not distinguish here, as Sen does, between a Social Welfare Function, which provides a complete transitive ordering or ranking, and a Social Decision Function, which need only be acyclical in respect to the complete set of alternatives. See Sen, *op. cit.*, Chapters 4 and 4*; cf. pp. 14–16. Nor do they consider the radical possibility of simply waiving any attempt to preserve a relation within every pair of alternatives that holds consistently over varying subsets; certainly procedures could be found, not committed to preserving such a relation, that 'will always produce a single preferred alternative' without assuredly providing 'a complete ranking'.

However, taking 'procedure' in the sense and perspective in which Barry and Rae are putting forward the argument, the argument is convincing enough; and instructive besides, in ways that they did not intend. For it completely overlooks the Revisionary Process and its virtues. (Of course, I do not claim more in virtue for the Process than that it sometimes produces an alternative preferred as an improvement over those previously contemplated.) Barry and Rae use 'procedure' in a sense for which the Revisionary Process is not a procedure; for them a procedure sorts out a given set of alternatives.

If they had envisaged procedures capable of adding to a given set an alternative designed to be preferred to all the alternatives initially given, would they have given any weight to the argument quoted? As regards such procedures, it is perfectly irrelevant to impose a condition that they should be 'capable of always providing a first choice from any list of alternatives'; or to make the point that if they were so capable there would have been no problem about ranking the set initially given.

Equally unmindful of the Revisionary Process is the classification that Barry and Rae give (p. 349 ff.) of overall responses to conflict of criteria: (1) forswearing any choice (2) ordering the criteria (3) allowing trade-offs. Surely inventing new alternatives is just as important *a priori* as any of these; yet they did not think of it when they were setting up the classicication. The third head – 'allowing trade-offs' – covers the possibility implicitly, though only in part. However, Barry's and Rae's explicit allowance for the possibility, already mentioned, comes in (p. 350) only very incongruously, under the first head, as a sort of concession that forswearing any choice may make a limited sort of sense if it is followed by expanding the set of alternatives.

[22] T. R. Burns and L. D. Meeker, in the paper that they presented to the Western Ontario conference, and in other writings with various collaborators, have criticized the current theory of collective choice from a point of view similar to the one taken in this paper and with similar results. Consider, for example, their stress upon the activity of players in transforming game situations until a situation with a satisfactory outcome is discovered (see, *inter alia,* W. Buckley, T. Burns and L. D. Meeker, 'Structural Resolutions of Collective Action Problems', *Behavioral Science,* Vol. XIX, No. 5, pp. 277–297, September 1974, (an offprint of which Burns kindly sent me). I was happy to find further support in several contributions to the Key Biscayne conference and in what I at least took to be the general trend of the discussion there. I have particularly in mind the recommendation by Plott (*op. cit.*) that we abandon the quest for global social orderings and remarks by Kenneth Shepsle on institutional arrangements for producing 'structure-induced equilibria' from limited agendas. Certain reservations that I have introduced in revising this paper take into account some observations made at the conference by Woody Kelly, which made it clear that limited agendas are not always helpfully limited, or helpfully augmented by limited moves.

JOHN CHAMBERLIN

A DIAGRAMMATIC EXPOSITION OF THE LOGIC OF COLLECTIVE ACTION[1]

1. Introduction

During the decade since the appearance of Mancur Olson's *The Logic of Collective Action,* a number of economists and political scientists have focused their efforts on further developing the model of group behavior analyzed by Olson.[2] While these efforts have contributed to both the precision and the generality of the theory of collective action, the theory remains a somewhat loosely connected set of verbal and graphical assertions about group behavior, and the central results of the theory have not all been demonstrated using the same analytical style. This paper attempts to remedy this deficiency in some major respects.[3] The analysis is made particularly easy to follow because of a result discussed in the Appendix which enables the analysis of both Cournot and Pareto optimal behavior using the same graph. It is hoped that this method of presentation will contribute to a consolidation of the theory of collective action and will also make the theory more easily accessible to the average reader.

The basic model of group behavior used by Olson is one of Cournot behavior in the presence of a public good. Two properties of such behavior which are of particular interest are the amount of the public good provided through Cournot behavior and the degree of suboptimality associated with such behavior in relation to Pareto optimal behavior. The relationships between these outcomes and variations in group size, wealth, intensity of preference and the price of the public good are examined below. The Pareto optimal outcome used in the analysis is the Lindahl equilibrium.[4]

2. The basic model

Consider a group of n identical individuals, each endowed with an identical budget (of amount w). There exist two goods, a private good Y with unitary price, and a pure public good X with price p. Figure 1 shows the allocation problem faced by the ith individual, whose equilibrium consumption is located at A. Provision of the public good by others, which results in an implicit

Hooker, Leach, and McClennen (eds.), Foundations and Applications of Decision Theory, Vol. II, 17–27. *All Rights Reserved*

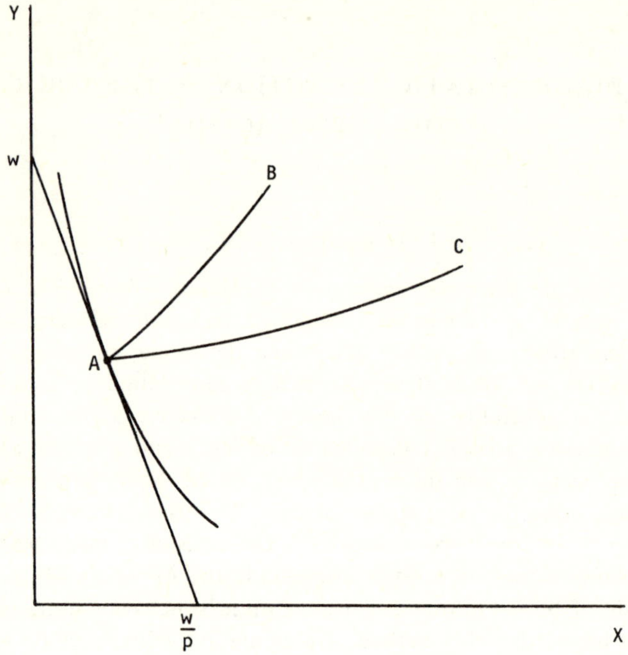

Fig. 1. Income–consumption and price–consumption curves.

increase in the income of the ith individual, will lead to changes in the individual's consumption as shown by the line AB, the income–consumption curve. Changes in the individual's consumption in response to decreases in the price of the public good (perhaps due to cost sharing arrangements among members of the group) are shown by the line AC, which is part of the individual's price–consumption curve (given the budget w).[5]

In Figure 2 a new budget line corresponding to a price of p/n for the public good has been added. It is shown in the Appendix that the amounts of the public good provided under Cournot and Lindahl behavior are those associated with the allocations E_1 and E_2 respectively. The Cournot equilibrium occurs where the income–consumption curve intersects the budget line (with price p/n for the public good), and the Lindahl equilibrium occurs where the price–consumption curve intersects this same budget line. The amounts provided at these equilibria are X_{E_1} and X_{E_2}, respectively.

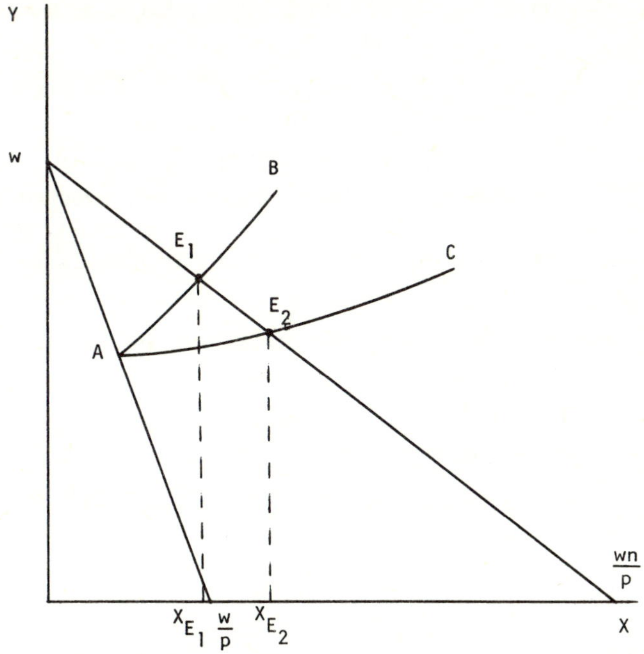

Fig. 2. Cournot and Lindahl equilibria.

3. THE COURNOT EQUILIBRIUM

The following properties of the Cournot equilibrium are obvious from Figure 2:

1. X_{E_1} is an increasing function of group size (n) and wealth (w).
2. X_{E_1} is a decreasing function of the price of the public good (p).

It can also be easily shown that:

3. X_{E_1} is an increasing function of the intensity of preference for the public good.[6]

With the exception of the relationship between group size and X_{E_1}, these conclusions are obvious to anyone familiar with the basic model of consumer choice.[7] They are listed here only for completeness and because the discussion below will consider the relationship between these variables and the degree of suboptimality associated with Cournot behavior, as measured by $(X_{E_2} - X_{E_1})$.

4. The Degree of Suboptimality

Since in Figure 2 the curve *AC* must always lie below curve *AB*, it is always true that X_{E_1} is less than X_{E_2}. That is, as Olson's original argument showed, Cournot behavior results in the suboptimal provision of a public good. The following conclusions concerning the degree of suboptimality are obvious from Figure 2:

4. the more price elastic the demand for the public good, the greater the degree of suboptimality $(X_{E_2} - X_{E_1})$,

5. the more income elastic the demand for the public good, the less the degree of suboptimality.

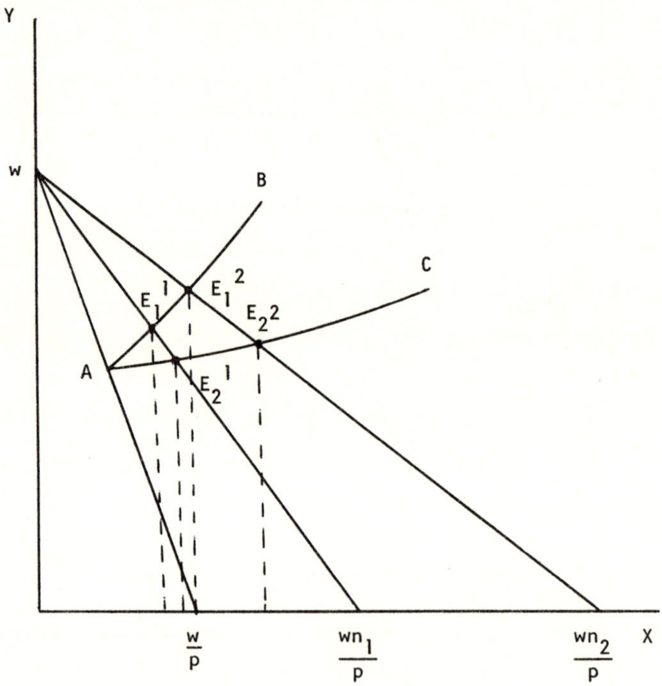

Fig. 3 Differences in group size.

4.1 *Group Size and Suboptimality*

Figure 3 shows the case of two groups of sizes n_1 and n_2, where $n_1 > n_2$. The four equilibria are shown, where the superscripts refer to the groups and the subscripts to the two types of equilibria. Another of Olson's central conclusions follows from Figure 3:

6. $(X_{E_2^1} - X_{E_1^1}) < (X_{E_2^2} - X_{E_1^2})$; that is, the larger the group, the more suboptimal the provision of the public good.

4.2. *Wealth and Suboptimality*

Figure 4 shows the case of two groups whose members are endowed with different levels of wealth, w_1 and w_2.

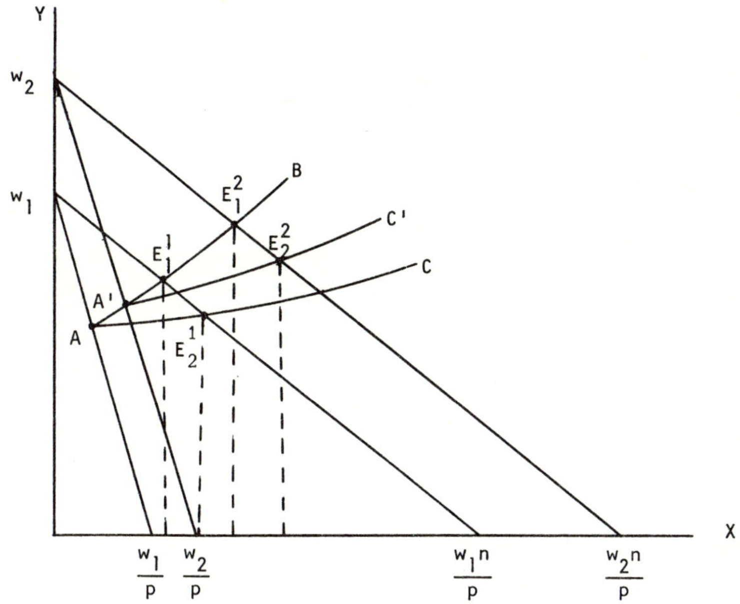

Fig. 4. Differences in wealth.

Since curve $A'C'$ may be drawn quite arbitrarily, subject only to the constraint that it lie between curves AC and $A'B$, it is impossible to draw any general conclusion about the relationship between wealth and the degree of suboptimality. For the case of Cobb–Douglas utility functions, the result is:

7. If the individuals have Cobb–Douglas utility functions, then the degree of suboptimality increases as wealth increases.

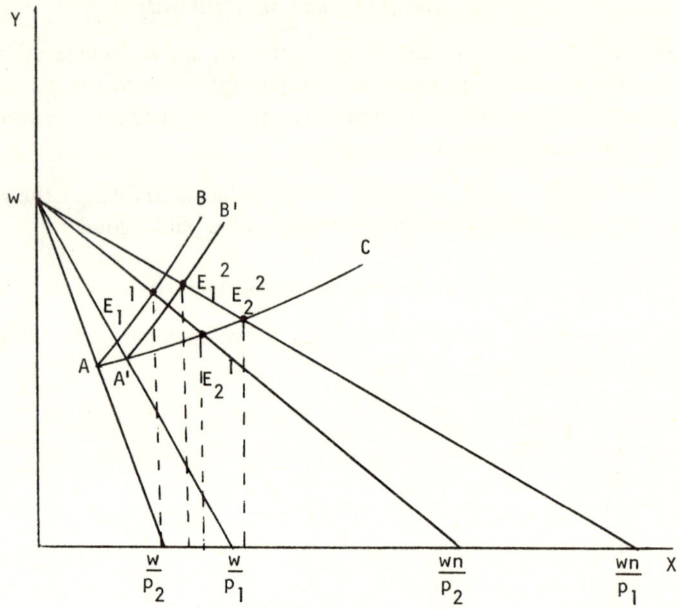

Fig. 5. Differences in Price

4.3. *Price and Suboptimality*

Figure 5 shows the case of two identical groups facing different prices p_1 and p_2, respectively, where $p_1 < p_2$.[8] It is impossible in this case to draw any general conclusions about how the degree of suboptimality is related to the price of the public good. The relationship depends upon the slopes of the two income-consumption curves *AB* and *A'B'* and upon the slope of the price-consumption curve. Since these slopes may take on quite arbitrary values (the only restriction being that the price-consumption curve lie below the income-consumption curve), there can be no general statement on this relationship. It is possible, however, to specify the relationship for a restricted set of individual preferences:

8. If the individuals have Cobb-Douglas utility functions, then degree of suboptimality associated with Cournot behaviour increases as price decreases.[9]

4.4. *Preference Intensity and Suboptimality*

Figure 6 shows the case of two groups of individuals with different intensities of preference for the public good. The more intense individuals have income-consumption and price-consumption curves labelled $A'B'$ and $A'C'$, respectively. As in the previous case, it is impossible to draw any general conclusion about the relationship between intensity of preference and the degree

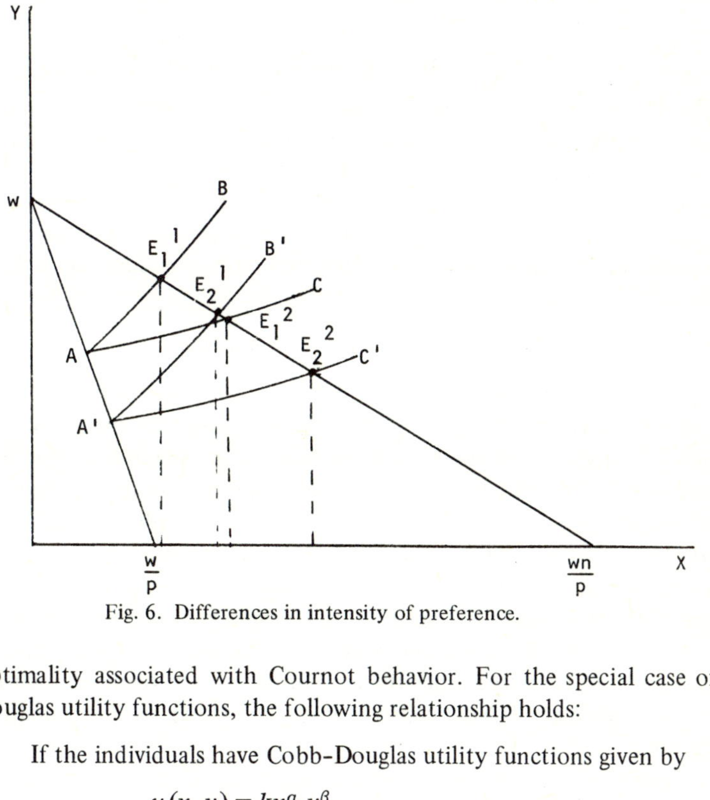

Fig. 6. Differences in intensity of preference.

of suboptimality associated with Cournot behavior. For the special case of Cobb-Douglas utility functions, the following relationship holds:

9. If the individuals have Cobb-Douglas utility functions given by

$$u(x, y) = kx^{\alpha} y^{\beta}$$

and an increase in the intensity of preference for x is indicated by an increase in the ratio (a/c), then the degree of suboptimality is an increasing function of intensity of preference for x if

$$\alpha/\beta < \sqrt{n}\,.$$

If a group is at all large, it is likely that this inequality will hold. Exceptions would be public goods of a 'life-or-death' nature.

5. Appendix

Consider again Figure 2. This appendix will demonstrate that E_1 and E_2 in Figure 2 are the Cournot and Lindahl equilibria, respectively, for the public goods problem under consideration. It is assumed that the n individuals have identical preferences, and identical budgets, and that neither good is an inferior good.

5.1. *Lindahl Equilibrium*

The Lindahl equilibrium in this case will be characterized by the n individuals each paying a price p/n for the public good (i.e., each bears an equal share of the burden). The optimal consumption for each individual at this price is found where the price-consumption curve intersects the budget line corresponding to a public good price of p/n. This is at E_2 in Figure 2, and the amount of the public good provided at the Lindahl equilibrium is X_{E_2}.

5.2. *Cournot Equilibrium*

The Cournot equilibrium is found most easily by examining the reaction curve of an individual, which is derived using the income-consumption curve.[10] Figure 7 shows such a reaction curve, labelled R. For the case of identical individuals, the Cournot equilibrium E is located at the intersection of the reaction curve, R, and a straight line through the origin with slope $(n-1)$.[11] At this equilibrium, each individual provides an amount X_i^E of the public good, and the total amount provided is nX_i^E. Figure 8 shows the reaction curve, now labelled R', transformed by changes in both axes. The equilibrium in Figure 8 occurs where the amountof the public good provided $(\Sigma X_i = nX_i^E)$ is (n/p) times the amount of money spent by each individual on the public good. Now consider Figure 9, which is a slight modification of Figure 2.

Note that if Figure 9 is rotated so that point D is viewed as the origin, the income-consumption curve is identical to the reaction curve in Figure 8 and that E_2 is the Cournot equilibrium, since the total amount of the public good $(\Sigma X_i = NX_i^E)$ is (n/p) times the amount of money (pX_i) spent by the individual on the public good, as in Figure 8.

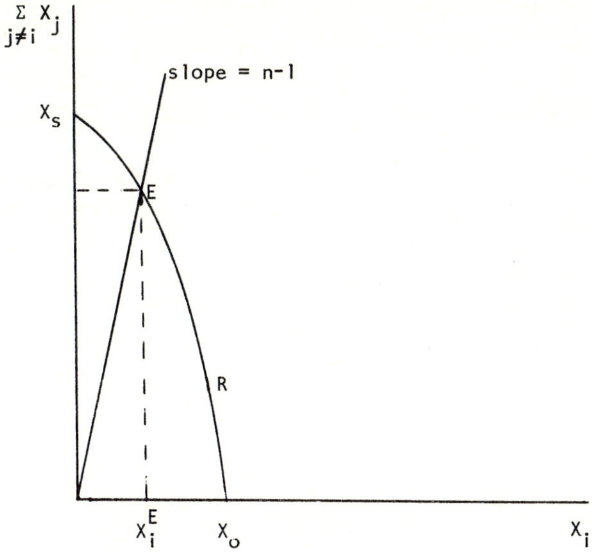

Fig. 7. Reaction curve and Cournot equilibrium.

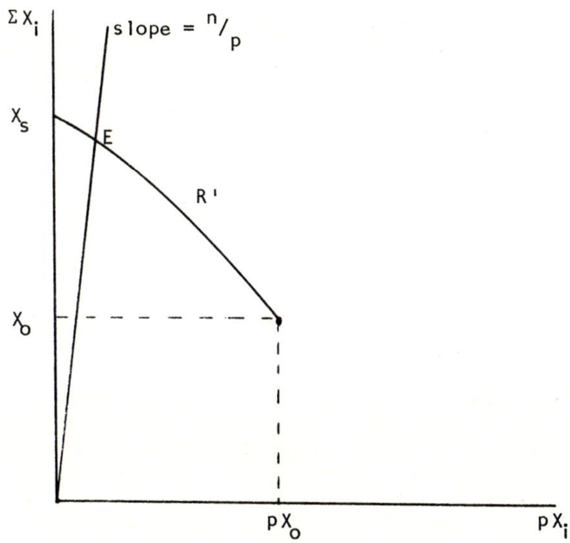

Fig. 8. Transformed reaction curve.

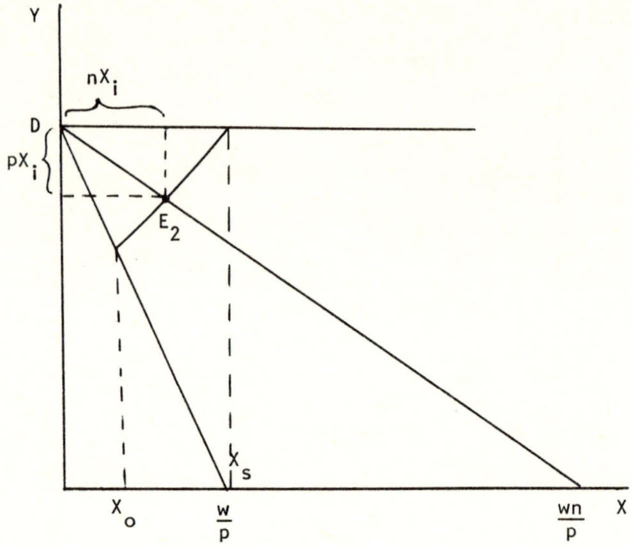

Fig. 9. The Cournot equilibrium and the budget line.

NOTES

1 I am indebted to Paul N. Courant for several helpful discussions during the preparation of this paper.

2 The most important works which use the style of analysis of the present paper (indifference curve analysis) are those of Chamberlin (1974), McGuire (1974), and Olson and Zeckhauser (1966).

3 The analysis here is restricted to a set of identical individuals. The analysis thus does not address the question of the equity of burden sharing discussed by Olson (1965) and Olson and Zeckhauser (1966), nor that of the properties of the outcome resulting from Cournot behavior among 'heterogeneous' individuals, as treated by McGuire (1974). Nor does this paper consider the provision of other than pure public goods.

4 See Musgrave (1959) for a discussion of the properties of the Lindahl equilibrium.

5 The analysis below is restricted to the case where neither good is an inferior good. This means that in Figure 1 *B* is above and to the right of *A* and that *C* is to the right of *A*.

6 The income–consumption and price–consumption curves for an individual with a stronger intensity of preference for the public good would lie below and to the right of those shown in Figure 2.

7 The result concerning the relationship between group size and X_{E_1} has been shown by Chamberlin (1974) and McGuire (1974).

8 In the analysis of interest group behavior, the price of the public good indicates the

rate at which individuals can transform private goods into effective political activity. Different prices facing two groups would be an indication of the fact that the group facing the lower price is more influential than the other group (can achieve the same change in the political outcome using fewer private resources).

[9] For a Cobb-Douglas utility function, the income-consumption curve is a straight line through the origin and the price-consumption curve is a horizontal line. It is possible that this conclusion can be shown valid for more general types of preferences. For instance, if preferences are homothetic (i.e. income-consumption curve is a straight line through the origin) and demand for the public good is price elastic and the price elasticity is constant, then it is true that the degree of suboptimality increases with price. I have been unable to obtain a more general specification of this relationship, however.

[10] See Olson and Zeckhauser (1966) or Chamberlin (1974) for details of deriving reaction curves.

[11] See Chamberlin (1974) for a fuller discussion of this procedure.

BIBLIOGRAPHY

Chamberlin, J. 'Provision of Collective Goods as a Function of Group Size', *American Political Science Review* 68 (1974), 707-716.

McGuire, M. 'Group Size, Group Homogeneity and Aggregate Provision of a Pure Public Good under Cournot Behavior', *Public Choice* 18 (1974), 107-126.

Musgrave, R. *The Theory of Public Finance,* McGraw-Hill, New York, 1959.

Olson, M. Jr., *The Logic of Collective Action,* Harvard University Press, Cambridge, 1965.

Olson, M. Jr., and Zechkauser, R. 'An Economic Theory of Alliances', *Review of Economics and Statistics* 48 (1966), 266-279.

LEON ELLSWORTH

DECISION-THEORETIC ANALYSIS OF RAWLS' ORIGINAL POSITION*

Bruno de Finetti, in an article[1] recently summarized in *Theory and Decision*, has compared the value judgments which should guide policy-making to the objective function of an optimization problem in Operations Research. He emphasizes that parochial features of current economic and social arrangements should not be presupposed in formulating the policy problem, and that the objective function is to depend not only on individual utilities but also on certain social interests concerning, e.g., equality and provision for future generations. Although de Finetti offers an attractive schema for approaching the policy problem that faces any society, he does not indicate, in the English summary anyway, how this ethical 'objective function' is to be arrived at. One approach compatible with de Finetti's general decision-theoretic framework for ethics is that developed by J. C. Harsanyi. Harsanyi views ethics as a branch of the general theory of rational behavior.[2] In particular, the 'objective function' for Harsanyian policy evaluation corresponds to the objectives that would be pursued by a fully rational, fully informed agent judging on the basis of 'impersonal social considerations' alone.[3] Ethics as rational, impartial choice is a common but vague philosophical notion; Harsanyi's accomplishment was to give content to 'impersonality' as a certain sort of risky choice: the agent chooses policy as if he had equal probability of being 'put in the place of' any individual in his society.[4] By appealing to the results of Von Neumann–Morgenstern utility theory, Harsanyi obtains a form of utilitarianism as his 'objective function'.

Utilitarianism has often been faulted particularly for making inadequate provision for the social interest in equality emphasized by de Finetti. Recently the anti-utilitarian moral philospher John Rawls has presented[5] a new variation of the Harsanyi-style argument that leads to a novel and attractively egalitarian (to some people) ethical principle, called by Rawls the difference principle. Rawls' theory is certainly worth examination by those interested in welfare economics, decision theory, and the policy sciences: already economists are applying the difference principle to problems of justice in taxation policy[6], for example.

Hooker, Leach, and McClennen (eds.), Foundations and Applications of Decision Theory, Vol. II. 29–45. *All Rights Reserved*

The difference between Rawls' approach and Harsanyi's is in the definition of 'impartiality.' In the argument-scheme of Sections 26–27 of *A Theory of Justice* (which I refer to as an 'ideal participant' argument), Rawls treats ethics as a rational decision made not under risk but uncertainty[7]. This change opens up a whole new range of possible decision rules besides the expected utility rule of Harsanyi. Elsewhere I have criticized Rawls' own arguments for the difference principle as the proper solution to this problem. In this paper I use certain axiomatic treatments of decision under uncertainty to analyze the problem formulated by Rawls. The analysis produces new insights, I believe, about the viability of alternate versions of the difference principle *vis-a-vis* utilitarianism, and reveals clearly the role of certain crucial postulates in the arguments for each, some of which have not been fully specified before.

In the ideal participant argument the choice of a principle of justice is made from the point of view of an arbitrary future participant in a society who does not know his own personal characteristics or role in society. One might imagine this participant as an amnesia victim who knows the general facts about his society and that he is a member – but not which one. This condition corresponds to Rawls' 'veil of ignorance'. Thus, the ideal participant approach reduces the choice of principles of justice to a problem of individual decision making under uncertainty, with the conditions of uncertainty specified to guarantee a 'fair' choice.

The structure of the decision problem under uncertainty facing the 'amnesia victim' is simple. His choice alternatives are competing practices for his society, and his choice will affect his own (unknown) future interests. He knows the utilities of the members of his society for each alternative, but is uncertain which member he is: the different members play the role of 'states of nature' in this decision problem. The amnesia victim applies some rule for individual rational decision under uncertainty to select the practice that seems best to him from his necessarily impartial viewpoint. The practice that would be selected is the 'just' one; and the *social analogue* of the individual decision rule used by the ideal participant can be taken as the principle of justice. If different principles might be applied to different cases, a 'mixed' rule with several clauses of the form 'in circumstances *C* use rule *R*' can be adopted.

Rawls compares the difference principle to the maximin principle for individual choice uncertainty. For in any choice among alternative practices, the difference principle selects the one best for the worst off 'representative man'. Such a choice would look like this to our ideal participant:

SITUATION I

	I_1	...	I_n
P_1	E_{11}	...	E_{1n}
⋮	⋮		⋮
P_m	E_{m1}	...	E_{mn}

Here the P_i are sets of alternative social institutions with different associated reward distributions, the I_j are the various 'representative men' in society (roughly, social classes), and E_{ij} is the average reward level from the ith alternative practice P_i to persons in the jth social class I_j. If our amnesia victim has to choose one of these practices for his society, he might use the *maximin* principle, i.e., choose an alternative whose worst payoff – payoff to the least advantaged class – is at least as large as the worst payoff of any other alternative. But this is precisely the choice dictated by the difference principle. Thus, the difference principle is the social analogue of the maximin rule for individual choice under uncertainty.

In similar fashion, the average utility rule, chief rival of the difference principle, is the social analogue of the so-called 'Laplacean' rule for individual choice under uncertainty. Consider Situation I again with a somewhat different interpretation: the amnesia victim is uncertain about which member of society he is, rather than merely which class he is in. Everything is as before except that I_j is the jth individual in society – the states of nature are the n members of society – and E_{ij} represents the utility to the jth individual of the ith practice (with interpersonal comparability[8] assumed). If the ideal participant resolved the modified Situation I by the Laplacean rule, he would assign equal probability to each state of nature and then maximize expected utility in the usual way. Clearly this produces the same social choice as prescribed by the average utility principle, for the veil of ignorance forces the ideal participant, in maximizing his own expected utility, to maximize average utility for society as a whole. Of course, it is unrealistic to assume that our amnesia victim could really have this much information. He should, however, be able at least to estimate the average welfare of each social class of society, and then can weight that utility according to the percentage of the population in each class, which produces the same result as the more detailed calculation above. It would certainly not be reasonable to treat the social classes themselves as equally probable for this calculation, since (as I have argued in the work cited) there is no reason for the veil of ignorance to deprive the decision maker of knowledge of relative class sizes, and he would know anyway that there are generally wide disparities in class size.

Even when information on relative class size is available, Rawls denies that the ideal participant should make use of it. This is because Rawls rejects the use of the principle of insufficient reason to determine probabilities in the original position. However, his reasons for this rejection are not sound, as I have shown elsewhere. I have also shown that the conditions Rawls proposes as favoring maximin are not present in the ideal participant choice properly construed.

To get a principle of justice from the ideal participant argument, we need to know whether the participant would use maximin or the Laplacean rule to resolve Situation I. If there were a generally accepted theory of rational decision under uncertainty, we could appeal to it to resolve the impasse. Unfortunately, there is not such theory. However, it *is* possible to characterize axiomatically all the major rules that have been proposed, including maximin and Laplace's. This has been done in similar ways by Chernoff and by Milnor[9], and the results have been summarized by Luce and Raiffa.[10] At least these axioms will allow us to carry the debate beyond a wrangle about the principle of insufficient reason, since the Laplacean rule is characterized by axioms that make no reference to probabilities. In the Milnor version, which I shall follow except on one point, it turns out that exactly one axiom holds for the Laplacean rule but excludes maximin, and another holds for maximin and excludes the Laplacean rule. Both rules are compatible with all eight other axioms proposed by Milnor for rational choice under uncertainty. Our task is to determine which of these two axioms is more appropriate to the sort of decision problem under uncertainty (d.p.u.u.) facing the ideal participant.

Maximin is characterized by the *column duplication* condition:

> If a d.p.u.u. is modified by deleting a repetitious column in the matrix representing the problem, then the ordering of alternatives produced by the decision rule is not changed.[11]

This is a characterization of 'complete ignorance' – if you have no clue about probabilities, then two columns (i.e., states of nature) that differ in name only can be conflated. E.g., consider the following d.p.u.u.:

EXAMPLE A

	Poor	Middle Class I	Middle Class 2	Rich
P_1	3	6	6	9
P_2	4	7	7	10
P_3	5	8	8	11

In this case, if you were really completely ignorant of the percentage of population in each class, Middle Class 1 and 2 might as well be treated as one class for decision making purposes (even if society distinguishes the two in ways unrelated to utility).

It is obvious that maximin satisfies column duplication, since maximin compares only the minimum elements of each row, regardless of the number of repetitions. This axiom eliminates the Laplacean rule, as shown by the following example:

EXAMPLE B

	I_1	I_2	I_3	Average Payoff
P_1	15	12	12	13
P_2	24	6	6	12

	I_1	I_2	Average Payoff
P_1	15	12	13.5
P_2	24	6	15

Here P_1 has the higher average in the first matrix, but when the repetitious column I_3 is deleted, P_2 becomes the Laplacean best action.

It should be clear now why Rawls need only consider the payoffs to 'representative men': because maximin satisifes column duplication, any more detailed specification of the 'states of nature' is superfluous anyhow, assuming that within classes utilities are indistinguishable for practical purposes. In what follows, though, Situation I interpreted so that the different individuals in society are the 'states of nature' is taken as the standard description of the choice situation facing the ideal participant. The advantage of this approach is that all relevant general information which might be available to the ideal participant (such as relative class size) is embodied in the matrix expressing the problem. The different decision rules are then specified axiomatically with reference to these standard matrices. If one only wants payoffs of 'representative men' to count, e.g., this can be accomplished by imposing the column duplication condition.

The column duplication condition corresponds very well to Rawls' 'thick' veil of ignorance which excludes the use of Laplacean probabilities. A somewhat thinner veil, characteristic of all rules for d.p.u.u.'s, merely specifies that states of nature must be treated impartially – two d.p.u.u.'s with matrices the same except for permutation of columns have the same ordering of alternatives (included in Milnor's 'symmetry' axiom[12]). While this axiom guarantees equiprobability of states when added to the other Laplacean-characterizing axioms, note that it does not mention probabilities.

The first important result of the axiomatic approach for us is the demonstration of a significant divergence between the difference principle and the lexicographic difference principle, which Rawls defines as follows:

> ... in a basic structure with *n* relevant representatives, first maximize the welfare of the worst off representative man; second, for equal welfare of the worst-off representative, maximize the welfare of the second worst off representative, and so on until the last case which is, for equal welfare of all the preceding *n*-1 representatives, maximize the welfare of the best-off representative man. We may think of this as the lexical difference principle. However, I shall always use the difference principle in the simpler form.[13]

The lexicographic difference principle *appears* to be a mere extension of the difference principle, which distinguishes some cases the other finds indifferent. In terms of Milnor's axioms, however, lexicographic maximin is not an 'extension' of simple maximin at all, since it violates the column duplication ('thick veil') condition:

EXAMPLE C

(i)

P_1	4	4	5
P_2	5	5	4

(ii)

P_1	4	5
P_2	5	4

Suppose we interpret lexicographic maximin for a comparison between two rows as requiring us to arrange each row's elements in increasing order and then to pick the first row with a payoff that exceeds the corresponding payoff in the other row, comparing left to right. Then in (i), P_2 is preferred by lexicographic maximin (henceforth LMM), and in (ii) P_1 and P_2 are indifferent, though by column duplication the ordering should be the same.

In terms of the 'thick veil' arguments Rawls uses, it makes sense for him to take maximin (henceforth MM) as primary instead of LMM, since it turns out that LMM does not satisfy the thick veil condition on which the rejection of Laplace's rule is based. However, it will be seen subsequently that the divergence between MM and LMM is not a mere technical curiosity, and that Rawls' confidence in the former is misplaced. MM suffers from fatal weaknesses not shared by LMM, which turns out to be a far more viable rule.

The Laplacean rule is characterized by Rubin's postulate. I give first the most intuitively evident form of the postulate, following Luce and Raiffa.[14] For reasons made clear below I shall refer to this axiom as the mixed strong independence postulate (MSIP):

> In a probability mixture of two decision problems under uncertainty with the same set of alternatives and states of nature, if the second d.p.u.u. has payoffs which do not depend upon the alternative chosen, then the ordering of the alternatives by the decision rule for the mixture problem should be the same as for the first d.p.u.u.

For example, suppose our amnesia victim were asked to resolve the following compound choice problem for his society:

EXAMPLE D

Case I
Probability = 0.75
That an oil shortage develops

	Poor	Rich
A_1 Pass standby oil export controls	40	60
A_2 No standby controls	35	70

Case II
Probability = 0.25
That no shortage develops

	Poor	Rich
A_1	50	100
A_2	50	100

Here the payoffs are interpersonally comparable estimates of average utility in each social class. For simplicity assume that the classes are equal in size and that their composition is the same in I and II – so that the 'fine structure' of the states of nature can be ignored. Rubin's postulate in its MSIP version says plausibly enough that the ideal participant should choose for this compound situation as he would in Case I alone. This is because in II it does not make any difference what he chooses. I shall give this postulate closer scrutiny later in the paper.

Rubin's postulate implies that if a constant payoff is added to each entry in a column of a d.p.u.u. matrix, the ordering of alternatives is unchanged (Milnor's 'column linearity' axiom). This seems reasonable. E.g., if a man is deciding which of two women to propose to, and if his grandfather suddenly offers $10,000 if the man's first-born child is male and named after Grandpa, and if the man believes his choice of spouse has no effect on the probability of having a male first-born – then there is no reason why Grandpa's offer should influence his decision. The column linearity condition is incompatible with both simple and lexicographic maximin as follows:

EXAMPLE E

(i)

P_1	5	9
P_2	1	10

(ii)

P_1	5+10	9
P_2	1+10	10

By column linearity, the preferred outcome should be the same in (i) and (ii), but both maximin rules select P_1 in (i) and P_2 in (ii).

The 'thick veil' and Rubin's postulates are plausible for conditions of complete ignorance, but between the two of them maximin and the Laplacean rule are both eliminated. Is there a rule which satisfies *both* axioms? Yes – the rule developed by Savage, which I shall refer to as minimax regret.[15] Naturally, a sacrifice has to be made, and the axiom that must be given up is Milnor's row adjunction axiom – an independence of irrelevant alternatives condition:

> The ordering between old alternatives is not changed by the addition of a new alternative.

Thus, for example, if a new alternative opens up suddenly for society, this won't make one of the formerly less preferred alternatives now *most* preferred in the expanded d.p.u.u. This axiom is extremely reasonable, and I have never seen a serious objection to it, as long as it can be assumed that introduction of a new alternative (new row in the d.p.u.u. matrix) does not affect probabilities of states of nature. From the point of view of rational choice by the ideal participant, it seems row adjunction should be accepted and minimax regret rejected.[16]

In the context of the 'post veil' arguments that Rawls introduces in Section 29, minimax regret fares far better, as I discuss elsewhere.[17]

There is a very fundamental objection to maximin that also applies to minimax regret. Consider the following d.p.u.u.:

EXAMPLE F

	I_1	I_2	I_3	I_4	I_5	I_6	I_7	I_8	I_9	I_{10}
P_1	100	100	100	100	100	100	100	100	100	10
P_2	10	10	10	10	10	10	10	10	10	100

In this situation the ideal participant must choose between social structure P_1, in which 90% are rich and 10% poor, and P_2 where the payoffs are the same but the proportions of rich and poor are reversed. According to maximin, the ideal participant must be *indifferent* between P_1 and P_2 – between a society where 90% are rich and 10% are poor and one in which the proportions are reversed! (It also happens that minimax regret gives the same counterintuitive result in this case – though, unlike MM, it would not give the same result for every permutation of the row elements).

The result can be considerably generalized. Arrow and Hurwicz[18] have shown that given conditions essentially similar to Milnor's column duplication plus the uncontroversial symmetry, row adjunction and a very mild dominance condition, the optimal alternative selected by the decision rule for d.p.u.u.'s can depend only on the maximum and minimum payoffs for each alternative. Any rule satisfying these conditions will thus fail to discriminate correctly between P_1 and P_2 in the above example – will be blind to relative class sizes. The root of the problem is clear: column duplication is at fault. The other axioms needed to derive this result cannot seriously be considered controversial as postulates for rational decision under uncertainty. We must, therefore, reject any decision rule that satisfies the column duplication axiom as inappropriate for the choices facing the ideal participant.

Column duplication, which corresponds as we have seen to Rawls' 'thick veil' of ignorance, has been persuasively touted as fundamental to characterization of choice under *complete* ignorance. What the present counterexample shows is that the ideal participant's Situation I problem cannot be construed as a case of complete ignorance, since it is perfectly clear that relative class sizes cannot be rationally ignored in cases like Example F. The complete ignorance condition only makes sense if one doesn't even know whether the different states of nature are all real possibilities. For suppose one knows that each state is a real possibility with an unspecifiable but non-zero probability. Then it makes no sense to conflate two identical columns when (assuming mutual exclusiveness) whatever the unknown probabilities p and q of these two columns, the probability of the combined state (viz., $p+q$) is greater than either. Now, this supposition is satisifed for the 'amnesia victim' who chooses in Situation I: he knows that each member of society has a genuinely existent set of values which could, with some non-zero chance[19] turn out to be his own. Therefore, for the ideal participant theory, we want not merely not to *require* the 'thick veil' of ignorance – we want to reject it as inappropriate and leading to unacceptable results.

The above considerations still leave lexicographic maximin and the Laplacean rule in the running. The importance of LMM's failure to satisfy column duplication should now be apparent: it saves LMM from erroneous decisions in cases like Example F.[20]

If an effective case can be made for Rubin's postulate, we have a practically decisive argument for the Laplacean decision rule, since only it satisfies this axiom plus all the remaining postulates (except for column duplication, of course, which has been eliminated). Before turning to a detailed examination of Rubin's postulate, let us consider what can be said if this axiom is not imposed. We will then not be able to rule out LMM conclusively on the basis of the axioms, and other rules are also no doubt possible. But there are still a number of considerations that militate against LMM. First, there is the problem that LMM shares with MM and, indeed, all lexicographic principles: a very small difference in one state between two alternatives may outweigh even the greatest countervailing differences in others.[21] For example, the case cited by Rawls as a potential counter-example to maximin[22] holds equally against LMM. It *is* extraordinary that very trifling gains to the least advantaged (who may not be bad off in absolute terms) should completely outweigh very large benefits to the better off. Second, LMM violates the rather plausible continuity axiom of Milnor. This cannot be considered sufficient grounds for rejection, however, since it is difficult to give continuity an intuitive significance (Milnor or Luce and Raiffa can be consulted for details). Third, and perhaps most important, LMM will give the same counter-intuitive result as MM for a case like Example F if we modify P_1 by decreasing the payoff to I_{10} by the slightest amount – both will select the 90% poor society P_2. LMM allows the interests of those other than the least advantaged to be considered, *ceteris paribus*, but the interests of the least advantaged still have an absolute priority over the rest.

The final question remains: how acceptable is Rubin's Postulate? In spite of the appeal of the postulate, which the previous discussion and example should have made clear, objections to the principle have been posed. E.g., Luce and Raiffa discuss three objections, but all of these either beg the question or have dubious significance. Luce and Raiffa give satisfactory replies to all three, and their book may be consulted for details.[23]

Even though the case for the Rubin axiom as presented by Luce and Raiffa seems quite convincing, to me at least, there is a consideration so far omitted. The axiom given by Luce and Raiffa is not quite the axiom used by Chernoff in his derivation of the Laplacean rule, and from which Milnor's column linearity follows. The real Rubin postulate is:

> If Q_1, Q_2, and Q_3 are three d.p.u.u.'s with the same alternatives and states of nature, and if for any alternative d and state of nature s, $u_3(d, s) = pu_1(d, s) + (1-p)u_2(s)$, where $0 < p \leq 1$, $u_i(d, s)$ is the utility in game i, $i = 1, 2, 3$, of the outcome of any alternative-state pair, and the utility function u_2 of Q_2 does not depend on the alternative selected (i.e., the matrix for Q_2 has all columns constant – possibly different constants for each column), then the ordering of the alternatives in Q_3 is the same as the ordering in Q_1.[24]

The axiom given by Luce and Raiffa and cited above is better referred to as the 'mixed strong independence postulate' (MSIP), and will be so called henceforth. To understand the difference between Rubin's postulate and MSIP, consider the following example:

EXAMPLE G

I

	I_1	I_2
A_1	10	2
A_2	0	12

II

	I_1	I_2
A_1	1000	500
A_2	1000	500

III[25]

	I_1	I_2
A_1	$\frac{1}{2}(10)+\frac{1}{2}(1000)$	$\frac{1}{2}(2)+\frac{1}{2}(500)$
A_2	$\frac{1}{2}(0)+\frac{1}{2}(1000)$	$\frac{1}{2}(12)+\frac{1}{2}(500)$

In III, each entry is the half–half probability mixture of the corresponding entires for I and II. Rubin's postulate requires that the ordering in III be the same as in I, where the role of Q_1 in Rubin's postulate is played by I, Q_2 by II, etc. Note that in the mixed problem III, referred to by Rubin's posutlate, *first* the uncertain state of nature and the chosen alternative single out a *lottery* as an outcome, *then* the lottery is played with known probabilities to determine whether the final outcome is the corresponding entry of I or II. As illustrated in Example D, however, MSIP refers to a different 'mixed' problem, $(\frac{1}{2}\mathrm{I}, \frac{1}{2}\mathrm{II})$, where *first* there is a lottery to determine which game is being played, *then* the uncertain state of nature along with the selected alternative determine the outcome. MSIP requires the ordering of alternatives for *this* mixed game to be the same as in I. The difference between the mixed game ($\frac{1}{2}$I, $\frac{1}{2}$II) and III is that in the former 'risk' preceeds 'uncertainty', but in the latter uncertainty preceeds risk. In each case, however, the outcome of a given choice of row combined with a given state of nature is either the corresponding entry of I or II with 50–50 odds.

It is so natural to take III and ($\frac{1}{2}$I, $\frac{1}{2}$II) as equivalent that Luce and Raiffa do so without even commenting on it. Nevertheless, this equivalence cannot be taken for granted.[26] An additional axiom is needed to establish the equivalence of MSIP and Rubin's postulate. The best way to see the need for an extra postulate is to compare MSIP with an analogous axiom for the theory of decision making under 'risk' (where objective probabilities for states of nature are given).

MSIP is so called by me because it is the analogue for mixed cases involving both risk and uncertainty of the strong independence postulate (SIP) for von Neumann–Morgenstern utility theory:[27]

> If lottery A is preferred to lottery B, then for any other lottery C, $(pA, (1-p)C)$ is preferred to $(pB, (1-p)C)$.[28]

Here, a lottery is any possible object of choice which gives certain definite prizes (which may be other lotteries or certain outcomes) with associated objective probabilities. $(pA, (1-p)C)$ is a lottery with two prizes, A with probability p and C with probability $(1-p)$, $0 \leq p \leq 1$. The strong independence postulate (SIP) says merely that if one prefers, e.g., \$1,000 to \$100, then one would prefer \$1,000 as a prize in a lottery with any other prize C to the same lottery with the same odds but with a \$100 prize instead of the \$1,000. For suitably restricted sorts of risky choices this axiom is quite evident and has wide acceptance as a principle of choice under risk.

Now, consider the mixed problem ($\frac{1}{2}$I, $\frac{1}{2}$II) of Example G again. Here we have a probability combination of two d.p.u.u's instead of two lotteries, but otherwise the situation is quite parallel to that described by SIP. Suppose, e.g., that I and II were *risky* choices instead of d.p.u.u.'s – each column now has an objective probability. Then each row of the matrices becomes a lottery. If in I A_1 is preferred to A_2, then by SIP A_1 is still preferred for the probability mixture of I and II (since the lotteries A_1 and A_2 are the same in II, playing the role of the constant lottery C in SIP). All that MSIP does is extend SIP to cases where rows of a d.p.u.u. take the place of the initial two lotteries A and B in SIP, and constant rows of another d.p.u.u. take the place of C.

Naturally, one could write a corresponding 'risk' version of Rubin's postulate.[29] Now, in axiomatic utility theory, where the postulates are uniformly concerned with choice among lotteries, there *is* an axiom that ensures the desired equivalence between SIP and the risk form of Rubin's postulate. What is required is a postulate for the reduction of compound lotteries:

> Any compound lottery is indifferent to a simple lottery with the same prizes, where the probabilities of each prize are calculated by the usual rules for combining probabilities.[30]

Thus, for example, the compound lottery $(\frac{1}{2}(1/3A, 2/3C), \frac{1}{2}A)$ is indifferent to the simple lottery $(2/3A, 1/3C)$. In Example G, if I and II were problems involving risk, so that in effect each row is a lottery, then it would follow by the reduction axiom for lotteries that choice A_i in the probability combination $(\frac{1}{2}\text{I}, \frac{1}{2}\text{II})$ is indifferent to choice A_i in III, $i = 1, 2$, so that the two problems would be the same.

Correspondingly, for the mixed case, if, in general, MSIP and Rubin's postulate are to be equivalent, then we need the following mixed reduction postulate (MRP):

> A mixed problem of the type described in MSIP has the same ordering of rows as the corresponding combined problem as described in Rubin's postulate.

I have taken a weak version of the reduction postulate sufficient to yield the desired result. The significance of this factorization of Rubin's postulate is twofold. First, MSIP is intuitively more transparent and convincing than Rubin's postulate. But, to make use of this clarity, it is necessary to specify the additional assumption that must be made to get Rubin's postulate. This does no *harm* to the argument, I think, since MRP appears to be quite unexceptionable intuitively. In the second place, however, there are views that would *accept* MSIP while *rejecting* Rubin's postulate. In short, there are views that do reject the mixed reduction postulate.

In particular, Daniel Ellsberg, in a very well-known article,[31] has put forward a view concerning decision making under uncertainty that seems to lead to such a position. Without going into the details, one can get some idea of the general position Ellsberg takes by reference to the decision rule that he offers as an example of the sort of rule he has in mind:

> For any decision problem, associate with each alternative A_i the index: $p \times \text{est}\,(A_i) + (1-p) \times \min\,(A_i)$, where est (A_i) is the

> expected utility of A_i given your best estimate of the probabilities of the states of nature (presumably using the principle of insufficient reason if you have *no* information). p is a parameter, expressing your degree of confidence in your estimate of probabilities (and ranging from 0 when your probabilities are mere unfounded guesses to 1 when you have well-established objective probabilities), and min (A_i) is the worst possible payoff you could receive from A_i. *Choose the alternative with the highest index.*[32]

Ellsberg's rule has the advantage of treating choices under complete uncertainty, risk, and everything in between by one uniform rule. It incorporates the scruples of those who, like Rawls, wish to discount Laplacean probabilities. Indeed, the rule reduces to maximin for decisions under complete ignorance, where p presumably goes to 0.

So Ellsberg's rule would lead to maximin in our d.p.u.u.'s, but it has the advantage of giving a way of to accept the highly intuitive MSIP without contradicting itself by accepting Rubin's postulate. Consider Example G again. Ellsberg's rule for this case, assuming that $p = 0$ for probability estimates of I_1 and I_2 and $p = 1$ for the lotteries, gives the maximin solution A_1 for I and A_2 for III – violating Rubin's postulate, since II has constant columns. Suppose we calculate the result of Ellsberg's rule for the mixed problem ($\frac{1}{2}$I, $\frac{1}{2}$II), however. It may not be clear how to resolve this problem by Ellsberg's rule, but the most reasonable approach seems to be to calculate the Ellsberg index for each row in the separate d.p.u.u.'s. I and II, then calculate the probability mixture of each row's indices in I and II for the choice under risk. Assuming again that $P = 0$ for the probabilities of I_1 and I_2, and also letting $p = 1$ for the more inclusive risk problem, we find that Ellsberg's rule will always resolve such a mixed case the same as case I – as required by MSIP. For each row of II is constant, so the index for each row in II is the same. So that the index for A_i in the *combined* game is $\frac{1}{2} \times$ (Index of A_i in I) $+ \frac{1}{2} \times$ (Index of A_i in II) which reduces to $\frac{1}{2} \times$ (Index of A_i in I) $+ K$, for all i. Clearly this transformation preserves the ordering of the indices in I.

I do not accept Ellsberg's rule, of course. I only use it to illustrate that Rubin's postulate depends on two more fundamental postulates, MSIP and MRP, and these two postulates do not necessarily stand or fall together, as the example shows.

In summary, we have seen that there are good reasons for rejecting principles that satisfy column duplication for d.p.u.u.'s and that maximin is eliminated as the solution to the original position. Lexicographic maximin is not eliminated, however. A conclusive argument for Laplace's rule over LMM follows from the acceptance of Rubin's postulate. Rubin's postulate was shown in turn to depend on two more fundamental axioms. Both of these axioms seem highly acceptable, and I do not know at present of any serious objections to either.[33] Nonetheless, it cannot be maintained that the rational preferability of average utility over the difference principle in the original position has been shown with finality, given the controversial and incomplete state of the theory of decision making under uncertainty. What can be claimed with some confidence is that the *simple* difference principle has been ruled out as a solution, and that the evidence so far indicates strongly that the average utility principle is the desired solution. Any justification of the difference principle as a theory of justice would probably have to proceed along very different lines.

University of Manitoba.

NOTES

[1] Bruno de Finetti, 'Utopia as a Necessary Presupposition for Every Significant Foundation of Economics', *Theory and Decision* **5** (1974), 335–342.

[2] John Harsanyi, 'A General Theory of Rational Behavior in Game Situations', *Econometrica* **34** (1966), 615.

[3] John Harsanyi, 'Cardinal Welfare, Individualistic Ethics and Interpersonal Comparisons of Utility', *Journal of Political Economy,* 1955, reprinted in E.S. Phelps (ed.), *Economic Justice*, pp. 266–285, Penguin, 1973.

[4] Ibid., p. 277, footnote 12.

[5] John Rawls, *A Theory of Justice* (henceforth TJ), The Belknap Press of Harvard University Press, Cambridge, 1971. Rawls regards other arguments he gives as more 'decisive' than the paradigm discussed here, but critical opinion has largely and I think rightly ignored these other arguments in favor of the argument of Section 26. These arguments are discussed at length in my 'Rawls' Argument for the Difference Principle', unpublished.

[6] See, e.g., Part Five of Phelps, *op. cit.*

[7] A choice involves 'uncertainty' if the outcome depends on contingencies for which no definite probabilities are known to the decision maker. If there are definite probabilities, the choice involves 'risk'. See R. D. Luce and H. Raiffa, *Games and Decisions* (henceforth GD), p. 13, John Wiley and Sons, New York, 1957.

[8] Although the average utility principle needs only 'unit comparability', 'full comparability' of utilities must be assumed in order to accomodate all possible rules. See A. K. Sen, *Collective Choice and Social Welfare*, p. 106, Holden-Day, Cambridge, 1970,

for definitions of these terms.

In the situation described, it is quite plausible that the amnesia victim, lacking knowledge of his personal utility function, would adopt the interpersonally comparable utilities of the various individuals as his own for decision purposes. The coherence of this procedure has been doubted, however (not cogently, in my opinion, but c.f., e.g., TJ, pp. 173–174).

Obviously the ideal participant paradigm requires an intersubjectively invariant evaluative standpoint for choice in the original position. Rawls' own solution via 'primary goods' faces serious difficulties, even were the explanation and relevance of primary goods clear. It seems that any solution should yield an 'index' with the properties of a utility function, since some outcomes of the ideal participants' choices are likely to involve objective risks. Given any cardinal 'index', the argument below goes through unchanged.

[9] Herman Chernoff, 'Rational Selection of Decision Functions', *Econometrica* **22**, (1954), 422–443. (Chernoff only derives the Laplacean rule, but his axioms are similar to and more general than Milnor's). John Milnor, 'Games Against Nature', reprinted as Chapter 7 of *Game Theory and Related Approaches to Social Behavior*, M. Shubik (ed.), Wiley, New York, 1964.

[10] GD, Chapter 13.

[11] Milnor assumes as one of his axioms that the decision rule for d.p.u.u.'s must produce a complete ordering of all alternatives.

[12] C.f. also GD, pp. 294–296.

[13] TJ, p. 83. I do not follow Rawls' abbreviation 'lexical' for 'lexicographic'.

[14] *Games and Decisions*, p. 290. I have modified the axiom to conform to Milnor's framework. Milnor himself assumes the less intuitively transparent 'column linearity' condition, which is implied by Rubin's postulate (see Chernoff, *op. cit.*, Theorem 2, p. 433).

[15] C.f. *Games and Decisions*, p. 280 ff. It should be noted that Savage discusses the social analogue of minimax regret – for rather different contexts than the present one – in Chapter 10 of *The Foundations of Statistics*, Dover, New York, 1972 (first published 1954).

[16] For further discussion see GD, pp. 281–282, 288–289.

[17] 'Minimax Regret and the Difference Principle', unpublished.

[18] 'An Optimality Criterion for Decision-making under Ignorance', in C. F. Carter and J. L. Ford (eds.), *Uncertainty and Expectations in Economics*, Basil Blackwell, Oxford 1972.

This result is also cited by Brian Barry in his discussion of the difference principle, *The Liberal Theory of Justice*, p. 91, Clarendon Press, Oxford, 1973.

Note that this general result does not apply to minimax regret (which does not satisfy row adjunction), though in particular cases like the example given similar counter-intuitive results are obtained.

[19] I do not want to insist on the 'unknown probabilities' interpretation of the uncertainty in the original position. I think that probabilities in the sense of Savage – subjective probabilities – exist in this situation, but they are not 'unknown' and anyway assuming them begs the question at issue. And it is not clear how we can assert that there are real but unknown 'objective' probabilities in this situation. But the point is clear

enough – the ideal participant will surely regard the real *possibility* of his being anyone in his society as sufficient to forbid column collapsing – and relative class size *will* make a difference to him. This is certainly obvious in the example given, where the max and min payoffs of each row are the *same*. Surely it would be absurd to argue that the vast differences in class sizes would not become decisive in this case.

[20] Another prima facie advantage of LMM can be mentioned here – it satisfies the weak dominance condition, which simple MM does not.

[21] This applies also to minimax regret. See Chernoff, *op. cit.*, p. 426.

[22] TJ, pp. 157–158. Rawls tries unsuccessfully to counter this objection to the difference principle. See 'Rawls' Argument for the Difference Principle'.

[23] GD., pp. 291–292, and references cited therein.

[24] Chernoff, *op. cit.*, p. 451. The axiom has been adapted to Milnor's format.

[25] This matrix reduces to

505	251
500	256

[26] Cf. the discussion of the independence postulate on p. 108 of Peter Fishburn, *Utility Theory for Decision Making*, Wiley, New York 1970.

[27] The analogy was pointed out by Daniel Ellsberg in 'Risk, Ambiguity, and the Savage Axioms', *Quarterly Journal of Economics* 75 (1961), 643.

[28] This corresponds to Luce and Raiffa's 'substitutibility' axiom in their treatment of utility theory, *op. cit.*, p. 27.

[29] Viz., let $A = (p_1A_1, \ldots, p_nA_n)$, $B = (p_1B_1, \ldots, p_nB_n)$, $C = (p_1C_1, \ldots, p_nC_n)$ be lotteries where the sequence of prizes in each lottery depends on the same states of nature with the same probabilities p_i, and the sum of the $p_i = 1$. Let $A' = (p_1(qA_1, (1-q)C_1), \ldots, p_n(qA_n, (1-q)C_n))$ and B' be defined correspondingly. Then if A is preferred to B, A' is preferred to B'.

SIP, on the other hand, asserts that $(qA, (1-q)C)$ would be preferred to $(qB, (1-q)C)$ if A if preferred to B.

[30] This corresponds to Axiom 2 of Luce and Raiffa's treatment, GD., p. 26.

[31] *op. cit.*

[32] See *ibid.*, p. 664.

[33] Other than examples where one has an urge to use a rule more conservative than Laplace's. E.g, Ellsberg, *op. cit.*, gives examples where there is a tendency to discount Laplacean probabilities. I am moderately optimistic that the 'right' answer can be adequately defended against most apparent counterexamples – see Raiffa's reply to Ellsberg in the sumposium, *ibid.*, pp. 690–694. I suspect the modification needed for residual difficulties (if any) in our cases would be a 'mixed' principle: the Laplacean rule if a utility 'floor' is guaranteed.

* This paper was prepared by invitation for the University of Western Ontario Colloquium on Foundations and Applications of Decision Theory, May, 1975.

DAVID GAUTHIER

THE SOCIAL CONTRACT: INDIVIDUAL DECISION OR COLLECTIVE BARGAIN?

1. In recent political thought the theory of rational choice has been deployed to afford new insight into the traditional theory of the social contract.[1] The theory of rational choice includes both decision theory, which formalizes independent choice by a single individual, and game theory, which formalizes interdependent choice by several individuals.[2] Each theory offers an interpretation of the social contract: the first, as an ideal individual decision; the second, as an ideal collective bargain. In this paper we outline and evaluate these two interpretations, and endeavour to demonstrate that only the second is defensible. We then consider briefly some of the weaknesses and implications of the contract theory, insofar as these are illuminated by the interpretation of the contract as a collective bargain.

2.

2.1. The social contract corresponds to no historical reality, and plays no role in the explanation of the origin and development of society. It is an analytical device for use in the assessment of social practices and institutions. The rationale for appealing to a hypothetical contract is found in the supposition that, what ever society actually is, it ought to be only an association for the mutual benefit of its members, so that its practices and institutions ought to be those which would result from an agreement which each would find rational to enter voluntarily.

We suppose that human beings are maximizers of individual advantage, or utility, which accrues to them primarily from the appropriation and use of material goods. In widely prevailing conditions of moderate scarcity, the total supply of these goods can be increased through cooperative action. Ideally, cooperation leads to an optimal state of affairs, in which the production and distribution of goods is such that no person can become better off except at the expense of some other person. But in cooperating to achieve such a state of affairs, maximizers of individul advantage face two problems. The first is that there are many possible optimal states of affairs, each consisting of one particular way of allocating the costs and distributing the benefits of cooperative action. The second is that each of these states of affairs can be

Hooker, Leach, and McClennan (eds.), Foundations and Applications of Decision Theory, Vol. II, 47–67. *All Rights Reserved*

realized only if at least some of the persons refrain from the direct maximization of their individual advantage. The outcome if everyone seeks to maximize his own surplus of benefits over costs is always sub-optimal.[3]

Fully effective cooperation is thus possible only if all persons correlate their actions in a manner which will require them to refrain from the direct pursuit of advantage. Since such correlation is in itself unstable, enforcement will be necessary. Hence all persons must accept some enforceable mode of correlating their actions. Since cooperation can benefit all of them, there must be at least one mode which is rational for all to accept, and which, therefore, could be the subject of voluntary agreement among them. We term such an agreement the social contract. Its particular content is established by solving the problem of rational choice for those persons party to it. We shall treat this problem abstractly, as the selection of a principle which, for each cooperative situation, determines a set of correlated actions for the cooperators on the basis of the expected costs and benefits involved. More concretely, the problem involves the selection of a set of social practices and institutions on the basis of their expected costs and benefits to the members of society. But we shall not consider here what these practices and institutions might be.

2.2. Adherence to the social contract, we have noted, is not always compatible with the direct pursuit of maximum individual utility. Not only does this raise the practical problem of enforcement, but also it raises the theoretical problem of the connection between the rationality of entering the contract and the rationality of adhering to the contract. There are foreshadowings of this problem in Plato[4], but the first direct discussion is provided by Hobbes.[5] I have examined the issue elsewhere, and endeavoured to show that from an initial identification of rationality with individual utility-maximization, one will rationally change one's conception of rationality so that one will consider it rational to adhere to an agreement such as the social contract despite the restraints on utility-maximization involved.[6] I shall not repeat those arguments here, but simply assume their conclusion. But it is important to the present enquiry to note that the rationality of adherence to the social contract must depend on the rationality of entering the contract, for this consideration will play an important role in our assessment of the two interpretations of the contract offered by decision and game theory. I shall argue that only if the contract is interpreted as an ideal collective bargain will every person consider that it would have been rational for him to have made such an agreement with every other person, and hence rational for him now to act as the agreement would have required.

3. The social contract is an exercise in hypothetical foresight. It represents the agreement one would have made, in an initial state of nature, in order to enter civil society. It does not represent the agreement one would now make, within civil society, in order to continue in this condition. Nor does it represent the agreement one would in hindsight have made, at the time of entry into civil society, could one have known the consequences. One must eliminate from consideration all information which one could have acquired only subsequently to entry into society, in determining what agreement would have been rational for one to make in the initial state of nature.

Thus the society which one is to enter by means of the contract can not be characterized in a determinate way. Its characteristics will be established, in large part, by the agreement which everyone makes. But prior to this agreement, one can know of the circumstances in which one will find oneself only that they will involve cooperative activity, and hence will include the practices and institutions on which it depends. And this knowledge is clearly insufficient to enable one to distinguish one's particular place and role in society.

If one were making an actual agreement, then one would know the circumstances prior to the agreement – the state of nature. But since the agreement is purely hypothetical, the state of nature which precedes the social contract must be as indeterminate as the society which follows from it. One thus can know of the circumstances in which one would find oneself prior to the social contract, only that they involve the absence of cooperative activity, and of all those practices and institutions on which it depends. And this knowledge again is insufficient to enable one to distinguish one's particular situation in the state of nature.

The social contract is thus the agreement a rational person would have made in the initial state of nature, in order to enter society, in ignorance both of his particular circumstances in the state of nature, and of his particular place in the society which results from the agreement. But some contractarian theories claim that there is a yet deeper level of ignorance affecting the agreement.[7] In their view, one must eliminate from consideration, not only all knowledge of one's particular circumstance, but also all knowledge of one's particular self. Since the social contract is purely hypothetical, the nature of the persons entering it must be an indeterminate as the circumstances of entry. One can thus know of oneself as a participant in the contract, only that one has certain abilities, and certain interests relating to the appropriation of material goods as well as other interests, and that one would use one's abilities to maximize the satisfaction of one's interests. But one can not know the nature of either one's abilities or one's interests. One can know that one has a utility-function, but not what that utility-function is.

Now we may not find the requirement convincing, that in the contract situation one not know who one is. Fortunately, we shall not have to decide this matter. For we shall find that the requirement is necessary only on the interpretation of the contract as an ideal individual decision, and as I have indicated, we shall reject this interpretation. On the interpretation of the contract as an ideal collective bargain, it is a matter of indifference whether the parties to the contract assume that in the contract situation they know who they are.

We shall therefore assume for the present that the parties to the contract do not know who they are, so that our argument will beg no questions against the decision-theoretic interpretation of the contract. Given this 'veil of ignorance', the foremost contemporary contractarian, John Rawls, is led to argue:[8]

> It is clear that since the differences among the parties are unknown to them and everyone is equally rational and similarly situated, each is convinced by the same arguments. Therefore we can view the choice in the original position from the standpoint of one person selected at random. . . . Thus there follows the very important consequence that the parties have no basis for bargaining in the usual sense.

Rawls maintains that the constraints which limit the information one can take into account, in determining what agreement one would rationally have made in the state of nature, require that the contract be interpreted as an ideal individual decision rather than as an ideal collective bargain. We shall maintain exactly the opposite position. We shall show that the limits on information require one to treat the contract as a bargain, both from the standpoint of the hypothetical person in the state of nature, assessing the rationality of entering the contract, and from the standpoint of the actual person in present society, assessing the rationality of adhering to the contract.

4

4.1. We turn now to the decision-theoretic interpretation of the social contract. We select an individual at random, and attempt to present an argument which would convince him, insofar as he is rational, to choose a particular principle for social or cooperative action. Since all persons are similarly placed, an argument rationally sufficient for one person is rationally sufficient for any person. Hence if we show that a randomly selected individual would choose a particular principle for cooperative action, we may conclude that every individual would choose this principle, so that it must be the content of the social contract.

Let our individual, then, consider each possible principle for cooperative action. (Since our treatment is purely abstract and formal, the practical impossibility of this supposition need not concern us.) For each principle, let him determine the expected utility of its adoption for each party to the contract. Then, using some interpersonally comparable measure of utility, let him determine the average of these expected utilities for each principle. Since he does not know what social role he will play, he must assume all possibilities equally likely, and so this average will represent the expected utility of the adoption of the principle for him, given the knowledge available to him in the state of nature. He should then select that principle, or indifferently from those principles, which affords the maximum average of the expected utilities of the parties to the contract.

The necessity of using an interpersonally comparable measure of utility may be considered an objection to this procedure, given the notorious difficulties with such comparisons. It might be thought that our individual could determine the expected utilities of adopting each principle for cooperative action, not by taking the standpoint of each part to the contract into account, but by considering all possible social roles from his own point of view. For each principle he would determine the expected utility, to himself, of each social role, were the principle to be adopted, and then simply take the average of these expected utilities as the overall expected utility of the adoption of the principle for him.

But this procedure, although it avoids the interpersonal comparison of utilities, violates the requirement that in the state of nature, no one can know who he is or what his utility-function is. And this requirement is necessary to the use of the decision-theoretic procedure. For if we were to suppose that our randomly selected individual knew his own utility-function, then we should have to conclude that he would rationally choose that principle for cooperative action which would afford him maximum expected utility. And we should then have no reason to suppose that every other person would rationally choose that same principle. Differing assessments of social roles would lead to differing evaluations of the consequences of adopting a particular principle, and hence to differing comparisons among such principles.

Provided, then, that we accept the interpersonal comparison of utilities, we seem to have a solution to the problem of the social contract. The contract should prescribe that principle for cooperative action which maximizes the average of the expected utilities of the parties to the contract; in other words, the contract should prescribe the average-utilitarian principle.[9]

But there is a fundamental objection to the argument for the averageutiliarian principle. In presenting that argument, we identified the expected utility of adopting the principle, for our randomly selected individual, with the average of the expected utilities of adopting the principle, for all of the parties to the contract. But this identification is inadmissible. The average of the expected utilities of different persons is not, and cannot properly be identified with, the average expected utility for a particular person. As Rawls insists, the average of expected utilities is not 'as expectations should be, founded on one system of aims'.[10]

In individual decision-making under uncertainty, one determines the probability-weighted average of one's utilities for the several possible outcomes. Insofar as one is rational, one selects that court of action which maximizes one's expected utility. If the actual outcome proves unfortunate, one will still consider one's decision to have been reasonable, because of its expected utility. If the decision involves an agreement, one will still identify with the making of the agreement, and hence one will consider it rational to adhere to the agreement even if non-adherence now affords greater expected utility.

Selecting the average-utilitarian principle for cooperative action is not an instance of rational individual decision-making under uncertainty. One's decision does not maximize one's expected utility, but only the average of the expected utilities of the members of society. If the actual outcome proves unfortunate for oneself, one will not consider one's decision to have been reasonable, because one will not view it as having maximized one's expected utility. The decision involves an agreement, but one will not identify with the making of the agreement, and hence one will not consider it rational to adhere to the agreement, if non-adherence affords one greater expected utility.

We may express this objection to the argument for the average-utilitarian principle in another way, by pointing out that the requirement that in the state of nature, one not know who one is, which must be accepted to suppose that the principle rationally chosen by one person would be rationally chosen by every person, also prevents identification of one's present self with the decision-maker in the hypothetical state of nature. Indeed, the decision-maker is not a person, for he has no utility-function – his decision is based on the averages of the expected utilities of different persons, each with a separate utility-function. In eliminating the differences between persons, which is necessary to treat the problem of the contract as an ideal individual decision, the individuality of the decision-maker has been eliminated, and this is incompatible with treating the problem of the contract as an ideal individual decision. The average-utilitarian principle rests on a contradiction.

There is, then, no reason for any person to consider the choice of the average-utilitarian principle in the state of nature as the choice he would have made. Since one does not identify oneself with the average utility of society, one does not identify oneself with what the principle maximizes. The attempt to show that a randomly selected individual would rationally choose to cooperate with others on the basis of average-utilitarianism fails.

4.2 Is it possible to avoid the contradiction in the reasoning leading to average-utilitiarianism, while retaining a decision-theoretic interpretation of the social contract? Is it possible to retain the requirement that in the state of nature one not know one's utility-function, and yet to found one's reasoning on a particular utility-function, so that the selection of a principle for social cooperation is a genuine instance of rational individual decision-making? We may interpret Rawls as offering a solution to the problem of the social contract which purports to satisfy just these conditions.[11]

Let us again select an individual at random. Since in the state of nature he does not know who he is, he must consider the possibility of being any of the parties to the contract. Now unless there is a mode of cooperation which is best for everyone – a mode which affords a unique optimal outcome, it is not possible to select a principle for cooperative action which would maximize the expected utility of each party to the contract. Hence, once society is established, our individual can not reasonably object to the contract simply because it fails to maximize his own expected utility.

Let us now suppose that our individual turns out to be peculiarly unfortunate – turns out to be the least-advantaged person in society. Now if the principle for cooperative action maximizes the expected utility of the least-advantaged person, our individual can have no reasonable complaint against its selection, for he must realize that no matter what principle was selected, someone would expect to be at least as badly off as he is. On the other hand, if the principle for cooperative action does not maximize the expected utility of the least-advantaged person, then our unfortunate individual could complain that had some alternative principle been selected, no one would expect to be as badly off as he is. Thus, it is argued, each party to the social contract, contemplating the possibility of being the least-advantaged person, will consider that in the state of nature he would rationally have selected that principle for cooperative action which would maximize the minimum level of expected utility for the parties to the contract.

The average-utilitarian attempt to solve the problem of the social contract failed because it treated the problem as one for individual decision, but did not provide an individual standpoint from which to make the decision. Rawls' solution provides such a standpoint, that of the least-advantaged person. Although no one in the state of nature knows that he is that unfortunate person, each knows that he might be least-advantaged, and this is sufficient to induce the selection of a particular principle for cooperative action. The social contract should prescribe the principle of maximin expected utility.

But the argument for the maximin principle is also open to a fundamental objection. In presenting the argument, we assumed that our randomly selected individual would identify with the standpoint of the least-advantaged person, in choosing a principle for cooperative action. But this identification is purely arbitrary. A person who finds that in society he is in a more advantaged position will not consider that the maximin principle was rationally selected, simply because it maximized the expected utility of someone else – someone other than the person who he turned out to be. Each party to the contract will insist that, although the decision in the state of nature must be made in ignorance of who he is, it must prove to be reasonable for him, no matter who he turns out to be. In other words, it must be reasonable from every standpoint, and not just that of the least-advantaged person.[12]

But is this possible? No principle for cooperative action can maximize everyone's expected utility. The attempt to select a principle which would be reasonable from every standpoint led to average-utilitarianism, a principle selected from no one's standpoint. The attempt to select a principle from some standpoint led to maximizing minimum utility, a principle selected from an arbitrary standpoint. Either the social contract must represent a decision made from some particular standpoint, or from some function of particular standpoints. If the former is arbitrary, and the latter does not provide a basis for individual decision-making, then the problem of the social contract can not be solved by interpreting the contract as an ideal individual decision.

5.

5.1. The game-theoretic interpretation of the social contract has no problem with the standpoint from which the principle for cooperative action is to be selected. On this view the contract is an ideal collective bargain, and a bargain is rational if and only if it is rational from the standpoint of each of the bargainers. It is therefore a matter of indifference whether or not each person in the state of nature is supposed to be able to know who he is. For each person knows that he is a party to the contract, and so, whether or not he can actually identify himself with a particular party, he knows that if the contract is a rational bargain, it is rational for him.

To interpret the contract as a bargain, we must first characterize the bargaining situation, in which a principle for cooperative action is to be selected. We must then show that there is a determinate solution to the bargaining problem in this situation. In other words, we must show that rational bargainers will agree on a particular principle for cooperative action. If we can show this, then we can identify this principle with the content of the social contract, and so successfully complete the interpretation of the contract as a bargain.

The bargainers are the parties to the contract. We identify each bargainer by a utility-function. This does not violate the requirement that no party to the contract know his utility-function, for we need not suppose that anyone in the state of nature knows which bargainer he is. To have a well-defined bargaining situation we must assign an expected utility for each bargainer to a *status quo* outcome, and to each of a set of possible bargaining outcomes. Each member of the set must of course afford each bargainer an expected utility greater than that afforded him by the *status quo*.

We identify the *status quo* with the state of nature situation. But since the circumstances of the state of nature are not known to the bargainers, the expected utility of the *status quo*, for each bargainer, can not be the utility of a determinate state of nature, but rather an average over the utilities of all possible states of nature, all possible sets of pre-social circumstances. (Again, the practical impossibility of this need not concern us here.)

The adoption of any principle for cooperative action affords each person an expected utility greater than that of the state of nature. Hence we may identify the possible bargaining outcomes with the expected outcomes of adopting the possible principles for cooperation. Now the circumstances of each possible society, as we may term the outcome of adopting a principle for cooperative action, are unknown to the bargainers. Hence the expected utility of a society for a particular person must be the probability-weighted average of the expected utilities of the various roles possible for him in that society.

In supposing that the parties to the social contract are all similarly situated, we have in effect assumed that no distinction can be made among persons, from the standpoint of the state of nature, in terms of the roles open to them in any possible society. But although this assumption is necessary to the decision-theoretic attempt to interpret the social contract, since it is necessary to ensure that all persons will reason to the same decision, it is not necessary to the game-theoretic attempt. We may suppose that we can not correlate each bargainer, identified by a utility-function, with any determinate set of circumstances in any society, but we need not suppose that we can not correlate different bargainers with different probability-distributions over possible social circumstances. The differences in the evaluation of a society by different persons need not reflect only their different utility-functions, but may also reflect the different probabilities of their being in social roles. This would represent a considerable weakening of the assumption that the parties to the contract find themselves in a veil of ignorance in the state of nature, but the interpretation of the social contract as a collective bargain does not require such a veil of ignorance.

Whatever decision we make about the availability of knowledge in the state of nature, we have now assigned an expected utility for each party to the social contract to the state of nature, as status quo, and to each possible society, as possible bargaining outcome. Hence we have characterized a well-defined bargaining situation.

5.2. We now select an individual at random, and attempt to present an argument which would convince him, insofar as he is rational, to agree to the adoption of any member of a particular set of principles for cooperative action. Since he can expect to benefit from society, we shall assume in our argument that this set will not be empty. A similar argument must be equally convincing to any other rational person. Hence our argument will enable us to correlate, with each party to the social contract, a set of principles for co-operative action, any member of which the particular party would rationally agree to adopt. We then consider the intersection of these sets. This comprises all those principles for cooperative action which everyone would rationally agree to adopt. Hence it must be the content of the social contract.

Let our individual first determine his expected utility for the *status quo*, i.e., the state of nature, which we represent as u^*. And let him determine his expected utility for the adoption of his most preferred principle, i.e., the bargaining outcome most favourable to himself, which we represent as $u^\#$. Now let him consider any other possible principle for cooperative action. Let his expected utility for its adoption be u; then u must be greater than u^* but less than $u^\#$.

The greatest gain which our individual could secure from society, in terms of expected utility, is $u^\# - u^*$. The gain which he would expect from adoption of the principle under consideration is $u - u^*$. Hence the ratio, $u - u^* / u^\# - u^*$, is a measure of the *relative benefit* which he could expect from the adoption of this principle, i.e., the benefit relative to the greatest benefit possible for him. Now we may suppose that the greater the relative benefit afforded him by the adoption of a principle, the more willing he must be to agree to its adoption.

Relative benefit is invariant with respect to positive linear transformations of utility-functions. Hence it is possible to make interpersonal comparisons of relative benefit, even without any interpersonal measure of utility. Our individual can therefore compare, not only the relative benefit for himself of of the adoption of different possible principles for cooperative action, but the relative benefits for all parties to the social contract of the adoption of these different principles.

Now suppose that our randomly selected individual considers that it would be rational for him to agree to the adoption of a principle if and only if he could expect a relative benefit no less than p. He must then agree that it would be rational for any other person to agree to the adoption of a principle if and only if that person could expect a relative benefit no less than p. There are now three cases to examine.

First, suppose that no principle for cooperative action is such that its adoption would afford everyone a relative benefit no less than p. Then rational agreement to select a principle is impossible. But everyone would benefit from the selection of some principle; hence rational agreement must be possible. Therefore our individual is mistaken in supposing that a relative benefit no less than p is necessary for it to be rational for him to agree to the adoption of a principle.

Second, suppose that some principle is such that its adoption would afford everyone a relative benefit greater than p. Then it is rational for some person to agree to the adoption of a principle, even though there is an alternative principle affording him a greater relative benefit, which would be rational for everyone else to agree to adopt. But this is not rational. Therefore, our individual is mistaken in supposing that a relative benefit as small as p is sufficient for it to be rational for him to agree to a principle.

Third, suppose that some principle is such that its adoption would afford everyone a relative benefit no less than p, and no principle is such that its adoption would afford everyone a relative benefit greater than p. The objections to the preceding cases do not arise. Rational agreement to select a principle is possible, but no one will rationally agree to the adoption of a principle, if there is an alternative affording him greater relative benefit, which would be rational for everyone else to adopt.

We supposed at the outset that each individual must consider the members of some set of principles rational for him to agree to adopt. We may now characterize the set as comprising those principles adoption of which would afford him a relative benefit no less than p, where p is the *maximin relative benefit*, i.e., the maximum minimum relative benefit afforded by any possible principle to all parties to the social contract.

It is therefore rational for every person to agree to the adoption of that principle, or any of those principles, which would afford each of them at least the maximin relative benefit. The set of principles affording maximin relative benefit to all is evidently the intersection of the sets of principles affording maximin relative benefit to particular individuals. This set constitutes the content of the social contract.

5.3. This solution to the problem of the social contract is an application of a general procedure for cooperative action which I have developed elsewhere.[13] And this procedure is itself an n-person generalization of a formal solution to the bargaining problem developed, quite independently, by Kalai and Smorodinsky.[14] If, in accordance with the usual assumptions of the bargaining problem, the possible bargaining outcomes constitute a closed, convex region in utility-space, then, as my discussion of the general procedure makes clear, maximin relative benefit must be maximum equal relative benefit. All of the parties to the social contract will rationally agree to adopt a principle for cooperative action which affords them the greatest possible equal expectation of relative benefit.

The Kalai/Smorodinsky solution is not the only one which has been proposed for the bargaining problem. The best known solution is undoubtedly that developed by Nash and anticipated by Zeuthen.[15] The essential difference between these solutions is that Kalai and Smorodinsky replace Nash's controversial 'independence of irrelevant alternatives' axiom with the requirement that if R and R' are two bargaining regions with the same *status quo* point (u^*, v^*), and if for any point (u, v) in R with $u \geqslant u^*$, there is a point (u, v') in R' with $v' \geqslant v$, and if (u_0, v_0) and (u'_0, v'_0) are the solutions for R and R' respectively, then $v'_0 \geqslant v_0$.

The rationale for the Kalai/Smorodinsky solution, and for my general procedure for rational cooperative action, is implicit in the argument of the preceding section. However, it may be useful to relate this rationale to the bargaining problem in a simple and straightforward way.

An alternative axiomatization of the Kalai/Smorodinsky solution, mathematically inferior but intuitively plausible, includes the usual requirements that the solution be invariant under positive linear transformations of utility-functions and that the solution be Pareto-optimal, and then imposes only the following additional requirement. Let (u^*, v^*) be the *status quo* point in the bargaining region R. Let $u^\#$ be the greatest utility in R which one person,

U, can receive, if V, the other person, receives at least v^*, and $v^\#$ be the greatest utility V can receive, if U receives at least u^*. Let the solution be (u_o, v_o); then if $u^* = v^*$ and $u^\# = v^\#$, $u_o = v_o$.[16]

This last axiom makes clear that the bargaining solution must depend strictly on two factors – the *status quo*, and the maximum each can receive compatibly with the other receiving the *status quo*. In the problem of the contract, it will be remembered, the first of these factors is the state of nature situation, and the second is the utility of each person's most preferred principle for cooperative action. The force of the axiom, together with the other requirements, is to ensure that everyone receives the same proportion of the difference between his *status quo* utility and his maximum utility, and that this proportion is as large as possible.

The rationale is evident. Each person knows that no one can expect a relative benefit greater than what he can expect. Each will demand this as a minimum condition, if he is to consider it rational to accept the bargain. But each will realize that his must also be a maximum condition, if others are to consider it rational to accept the bargain. Hence a bargain is rational if and only if it maximizes equal relative benefit for those party to it.

5.4. The interpretation of the social contract as an ideal collective bargain does not give rise to the objections that we raised against attempts to interpret the contract as an ideal individual decision. It is worth noting in the first place that the vexed problem of interpersonal utility comparisons does not arise if the contract is viewed as a bargain. But more important, the contract, as a bargain, is rational from the standpoint of every party to it. Every present member of society can identify himself with one of the parties to the contract in the state of nature, and can recognize that it would have been rational for him to have agreed to the adoption of the principle of maximin relative benefit. And this, I should argue, provides every present member of society with a basis for assessing the actual institutions of society, and for assessing the social actions of himself and others.

Let us suppose that the defenders of the average-utilitarian principle were to convince us that their principle was rationally chosen from an 'average' standpoint – rather than not being based on any individual standpoint, as we have claimed. Or let us suppose that the defenders of the maximin utility principle were to convince us that the standpoint of the least advantaged person was not determined purely arbitrarily, but that there was good reason to choose it rather than any other individual standpoint. We should still have no reason to interpret the social contract as an individual decision rather than as a bargain. For neither of these views permits every present member of society to identify with the decision-maker in the state of nature. Each requires that the parties to the contract, in the state of nature, not know who they are.

We can recognize the necessity of this requirement, if the contract must be interpreted as an individual decision. If there is no alternative way to make the contract an exercise in rational choice, then we must agree that the rationality of the contract can not be established from the point of view of each member of society, but only from some other point of view which they may be persuaded to accept. And we may then wonder whether it is plausible to use the contract as an analytical device in the assessment of social practices and institutions.

But of course, there is an alternative way to make the contract an exercise in rational choice, and on this alternative, the rationality of the contract is established from the point of view of each member of society. And this alternative is also the most natural way of interpreting the contract – for the contract, surely, is a bargain. It is an agreement among persons who find mutual advantage in cooperation, to determine the terms of that cooperation.

Let us conclude this section by providing our final characterization of the social contract. Let P be the set of all basic social institutions and practices – of those procedures which allocate social costs and distribute social benefits. Heretofore, we have identified each such set with a single principle for cooperative action, but we may now remove this over-simplification.

For any set of persons N, there will be a proper subset of P, $P(N)$, which includes all sets of basic social institutions and practices which are possible for N. $P(N)$ will be the set of possible societies, or better, the set of possible bases for societies, for the members of N. Then for each person i in N, there will be a subset of $P(N)$, $A_i[P(N)]$, which includes all sets of basic social institutions and practices which it would be rational for him to accept in order to enter society.

Then the social contract for N, $S(N)$, is simply the intersection of the sets $A_i[P(N)]$ for all i in N. It is the set of sets of basic social institutions and practices which it would be rational for every member of N to accept in order to enter society. If it is empty, then social cooperation can not be mutually advantageous for all of the members of N, and they can not constitute a rationally-grounded society. If it is non-empty, then a rationally-grounded society is possible for all members of N.

6.

6.1. There are at least two significant theoretical objections to the use of the social contract in assessing social institutions and practices. There are also, of course, evident practical objections to the possibility of determining the expected benefits of the state of nature, and the expected benefits of the adoption of each possible principle for cooperative action, for all parties to the contract. But our present concern is not practical. Before worrying about such matters, we must face the claim that the very idea of the social contract involves conceptions which make it inapplicable to the assessment of society.

The first of these concerns the *scope* of the contract. We have reasoned as if the set of persons party to the contract had its membership fixed. Now we can take for granted the present existence of human beings in societies, corresponding more or less closely to existing nation-states, and then attempt to assess each as if it were contractually based. This is in effect what Rawls does.[17] But although this division into nations can be explained, being by no means historically arbitrary, yet if our concern is with the assessment of social institutions and practices, then a division of persons into societies is itself a subject for assessment. Normatively, the existing division would seem to be an arbitrary one.

No doubt the members of most existing human groups are all better off than they would be were they in the non-social state of nature. Insofar as this is the case, the members have a basis for co-operative arrangements among themselves, which we have identified with society. It may then be argued that this justifies the existing national divisions. But clearly this is not so. For there are many alternative ways of partitioning mankind into groups such that everyone would be better off than in the state of nature. What we require is a justification of some particular division, which would not serve equally well as a justification of other, incompatible divisions.

It might then be argued that we should consider mankind as a whole to constitute the society of the human race. Included among the possible social institutions and practices from which a selection must be made, would be partitions dividing this society into sub-societies, or nations. The social contract would then be an agreement to adopt that principle for cooperative action which led to those institutions, practices, and nations, the existence of which would maximize the minimum level of expected relative benefit, for all persons.

One might object to this idea of the society of the human race, on the ground that some persons could expect to be better off in the state of nature, than in such a society. But I see no reason to accept this objection, and so shall assume that everyone would expect to benefit from a universal society. Nevertheless, there is a fundamental objection to the proposal. For it may well be possible to partition mankind into groups, such that for any two such groups *A* and *B*, every member of *A* is better off if the members of *A* form a society, but leave their relations with the members of *B* in the state of nature, than if the members of $A+B$ form a society, whereas every member of *B* is better off if the members of $A+B$ form a society, than if the member of *B* form a society, leaving their relations with the members of *A* in the state of nature.

Consider a plausible example of such a partition. Suppose that mankind is divided into two groups, the inhabitants of developed countries, and the inhabitants of less-developed countries. Then it is plausible to argue that every person in the developed countries would be better off if the developed countries constituted a contractually-based society, leaving their relations with the rest of mankind in the state of nature, than if there were a single society of the human race. And it is even more plausible to argue that every person in the less-developed countries would be better off if there were a single contractually-based society of the human race, than if the less developed countries constituted a society, leaving their relations with the rest of mankind in the state of nature.

The problem of determining the scope of the contract should now begin to appear. Suppose that it is possible to partition mankind into nations, in such a way that the members of some, but not all, prefer to belong only to their national societies, and not also to a society of the human race. Then those who prefer only national membership will insist that in the state of nature, they would rationally have entered a contract, not with all other persons, but only with members of their nation. They will therefore claim that the scope of the social contract, or rather of the social contracts, must be determined by the partition. On the other hand, those who prefer membership in a society of the human race will insist that, since all persons prefer such a society to the state of nature, all would rationally have entered a single contract, which may create nations but only within the framework of an overall society.

The first of these arguments is correct. It is a necessary, but certainly not a sufficient, condition for rational agreement among a group of persons, that each prefers agreement to no agreement. But it is also a necessary condition for rational agreement that there is no sub-group, each of whom prefers agreement only with other members of the sub-group, to agreement with all. It is not rational for persons to extend their agreement more widely than benefits them.

It is no objection to this position that in the state of nature, no one can know what his social circumstances will be, and so no one can know whether or not he belongs to a sub-group, the members of which prefer agreement only among themselves. For the social contract is a device for the assessment of social institutions and practices among persons who find it rational to co-operate voluntarily. The scope of the contract must be determined before the hypothetical state of nature is introduced; those in that state of nature are those who find it rational to agree among themselves. The state of nature is used in determining the content of their agreement, not in determining who shall be party to it.

But then how do we determine the scope of the social contract or contracts? There is no reason to suppose that there is a unique way of partitioning mankind into nations, no two of which can agree to constitute a larger society. Rather, there may be many, mutually exclusive partitions, each of which would seem to some persons to be the best basis for determining the scope of the social contracts. To decide among possible partitions, we require a theory of rational coalition formation. Whether such a theory is even possible raises issues which I can not begin to consider here.

6.2. The second problem which requires brief mention concerns the *time of entry* into the social contract. We have reasoned as if the contract were a simultaneous agreement among all persons. But there is no single time of entry, and indeed, there is no time at which all members of a society co-exist together.

Nevertheless, it may be argued that since no one in the state of nature can know his social circumstances, no one can know his circumstances of entry into the society. Hence for the purpose of the contract, we may suppose that all persons, past, present, and future, agree to the adoption of a principle for cooperative action which maximizes minimum relative benefit over time. Our ignorance of future persons clearly poses practical problems in determining the content of such a principle, but we have waived such issues here.

The first objection to this proposal is that it would seem to rule out the possibility of social progress. Given our conception of man as seeking to maximize his appropriation and use of material goods, we shall understand social progress as the per capita increase in the availability of such goods. Now we increase the availability of material goods, over time, by investing a portion of present production as capital for future production, rather than using it for present consumption. But this diminishes the quantity of goods available to earlier persons, in relation to the quantity which would be available to them if there were no progress, and per capita availability remained constant. Hence, since utility is a function of the quantity of available goods, those who save to provide capital for future production would increase their utility by consuming instead. Since saving makes more goods available in the future, it is a transfer from those with fewer goods to those with more. And this is incompatible with the principle of maximin relative benefit.

But this objection is unsound. We must remember that one's relative benefit is determined by one's position in the state of nature, and one's position in that bargaining outcome most favourable to oneself. Now the most favourable outcome for those in the first generation is that in which they consume all that they produce by cooperation. But the most favourable outcome for those in each succeeding generation is that in which they consume all that they produce by cooperation, given that those in each preceding generation have saved all their production in excess of that which would have been possible for them in the state of nature. Thus each succeeding generation can expect more goods, in its most favoured outcome, than the preceding generation could expect, in the outcome most favoured by it. And so the quantity of goods required if relative benefit is to remain constant, must also increase with each generation. Each generation must therefore save at that rate which keeps relative benefit constant over time at the maximum level. This is what is required by the principle of maximin relative benefit.

But now a second objection appears. Suppose that we partition people by generations. Then, although the members of each generation would wish to contract with all previous generations, they would not wish to contract with any future generations. It follows that there must be a separate contract for each generation. There will be no saving, but each will appropriate the maximum possible for its own use, by cooperation among its members, out of the stock which previous generations were simply unable to appropriate.

This is clearly an unsatisfactory model of society. If the theory of the social contract leads us to conclude that any two generations are in the state of nature with respect to each other, then we may well suppose that it is not a useless, but a pernicious theory.

But let us not assume such a conclusion hastily. Generations do not succeed each other, one marching off the stage as the next marches on. Although there is no time at which all members of a society coexist together, any two members are related by the ancestral of the coexistence relation: A coexists with N_1 who coexists with N_2 who . . . coexists with N_n who coexists with B. And this constant and complex overlapping provides us with a more satisfactory way to deal with time of entry problems.

Simultaneous time of entry to an agreement is not necessary. Partnership arrangements, as in a firm, provide for partners leaving and entering; eventually none of the original partners remains yet the partnership continues. Suppose we think of each party to the social contract as agreeing with all those with whom he coexists to the adoption of a principle for cooperative action which maximizes minimum relative benefit for all these persons. Now if we may assume, as in **5.3.**, that maximization of minimum relative benefit is maximization of equal relative benefit, then it will be evident that all parties to the contract must agree to the adoption of that principle for cooperative action which maximizes equal relative benefit for all. And this, it will be remembered, requires savings, so long as capital accumulation increases production.

Furthermore, we may suppose that in a world with finite material resources, at some point conservation measures will be needed to prevent a decrease in the per capita availability of material goods over time. Social progress, as we defined it in terms of the increasing availability of material goods, may no longer be possible, yet capital accumulation, and so saving, will be needed to maintain constant production, which under these circumstances will secure constant relative benefit.

Although our discussion has been very sketchy, we may conclude that our interpretation of the social contract as an ideal collective bargain seems to enable us to meet time of entry problems. We have defended the principle of maximin relative benefit, and with it the very idea of the contract, as applicable to the relations between all members of society, past, present, and future, and we have defended the necessity for saving, in terms of the principle of maximin relative benefit.

7. We conclude with a brief comment on the underlying individualistic ideology of the theory of the social contract. To suppose that the rationale for a principle for cooperative action is to be found in a hypothetical contract is to employ a strongly individualistic model for assessing society. It is to treat society as conventional, not natural, as an instrument to individual ends, not as a determinant of those ends.

One might suppose that to interpret the contract as an individual decision would be more consonant with the individualism of the contract theory, than to interpret it as a collective bargain. But this is not so. To understand the contract as an individual decision is in fact to obscure the strong individualism underlying the theory. This individualism is fully revealed only when the contract is understood as a bargain.

If the contract is considered an individual decision, then each contractor must be so placed that all make the same decision. Each must select the same principle for cooperative action. As we have seen, if each reasons from his own point of view, then not all will make the same decision. A common basis for decision must therefore be found, and this undercuts the individualism of the theory. We have examined both the attempt to find such a basis in the average of all individual standpoints, and the attempt to find such a basis in the standpoint of the least-advantaged individual. Whichever alternative one adopts, the result is a purported justification of the contract which does not rest on the interests of the individual contractor.

To consider the contract as an individual decision based on the average of all standpoints, is to impose a social constraint on the pursuit of individual interest in society. On this view, each may benefit only on terms which increase average benefit. To consider the contract as an individual decision based on the standpoint of the least-advantaged person, is to impose a strong egalitarian constraint on the pursuit of individual interest in society. On this view, each may benefit only on terms which maximize the minimum benefit. Now either, or both, of these constraints may appeal to us for all sorts of reasons – or perhaps, all sorts of feelings. But neither constraint is compatible with the theory of the social contract, properly understood.

If the contract is considered as a collective bargain, then it is based on the interests of each of the contractors. Its justification is in terms of the relation of each interest to all of the other interests. This is to impose only a weak egalitarian constraint on the pursuit of individual interest in society. On this view, each may benefit only on terms which permit proportionately equal benefits to others. But such equality permits great inequality in actual benefits, provided these rest on differences, either in the positions of the contractors in the state of nature, or in the positions available to them from all of the possible modes of cooperative action.

The social contract determines a principle for cooperative action. Rawls thinks of it as determining the principles of justice. We may generalize from this, and suppose that the mode of cooperation determined by the contract may be termed a morality. The aim, conscious or unconscious, of the dominant trend of modern moral philosophy has been to show that morality is to be understood as conventional, not natural, and that morality rests on the agreement of rational, self-interested men who are not, in themselves, moral beings. Given both our ordinary and our scientific ways of thinking about and understanding ourselves, this is a very plausible way of considering morality. Its result, I have argued, is to conceive of the principles of morality as the outcome of a bargain. And this result, if we take economics to be the fundamental science of human activity, is surely what we should expect.

University of Toronto

NOTES

[1] John Rawls, *A Theory of Justice*, Cambridge, Mass., 1971, is the prime example.

[2] The theory of rational choice also includes the theory of collective choice, which formalizes choice by a group of individuals. But it is of no relevance here, since the social contract precedes formation of a group.

[3] Sub-optimality results because individual maximization does not secure public goods, which provide individual benefit without compensating costs, or avoid public evils, which impose individual costs without compensating benefits. There is pollution but no sanitation.

[4] *Republic*, Book Two, in which Glaucon and Adeimantus put the case for injustice to Socrates.

[5] *Leviathan*, Chapter 15, in which Hobbes discusses the Fool's rejection of justice.

[6] 'Reason and Maximization', *Canadian Journal of Philosophy*, IV, 3, March 1975, especially pp. 427–432.

[7] Cf. Rawls, *op. cit.*, pp. 136–142.

[8] *Ibid.*, p. 139.

[9] Average-utilitarianism prescribes maximization of average happiness (or utility), in contrast with classical utilitarianism, which prescribes maximization of total happiness. Cf. Rawls, *op. cit.*, pp. 161–162.

[10] *Ibid.*, p. 175.

[11] I am not reproducing Rawls' own position, which prescribes maximization of the minimum level of primary goods, not of utility; I am also not reproducing Rawls' arguments for his position. This is an interpretation, a 'Rawls-type' position.

[12] One might attempt to defend Rawls by insisting that only the principle of maximin expected utility is reasonable from the standpoint of the least-advantaged person; hence no other principle can be reasonable from every standpoint. But in my paper, 'Justice and Natural Endowment: Toward a Critique of Rawls' Ideological Framework', *Social Theory and Practice*, pp. 3–26, I explore this approach and show that what results is not maximin expected utility, but the principle which I develop in Section 5 of this paper, interpreting the contract as a collective bargain.

[13] 'Rational Cooperation', *Nous* 8 (1974), 53–65.

[14] E. Kalai and M. Smorodinsky, 'Other Solutions to Nash's Bargaining Problem', *Econometrica* **43** (1975), 513–518.

[15] J. Nash, 'The Bargaining Problem', *Econometrica* **18** (1950), 155–162; F. Zeuthen, *Problems of Monopoly and Economic Warfare*, pp. 111-121, London, 1930.

[16] Cf. my unpublished paper, 'Rationality, Fairness, and the Bargaining Problem' for a full development of this axiomatization.

[17] Rawls, *op. cit.*, pp. 7–8, 377–378.

ALFRED KUHN

ON RELATING INDIVIDUAL AND SOCIAL DECISIONS

1. Introduction

By intuition or design of its planners there is a certain unity among the parts of this symposium. However, as a participant-observer I also felt a significant disjunction. Hence in saying my own piece I will try to emphasize some problems of relationships of parts. The particular framework used in this paper is a system-based approach to unified social science, as found in more detail in three books (Kuhn, 1963, 1974, and 1975), which in turn reflect a core of the large literature in system analysis.

Some elements from system analysis are useful for this purpose, in two respects. The first is to distinguish the kind of systems for which the term 'decision' makes sense from those for which it does not – which topic will be discussed immediately below in this Introduction. The second is to note that the very structure of the former kind of system, as seen by the system analyst, can provide a useful context for analyzing decisions and for clarifying the meaning of both 'causal' and 'behavioral' explanations of purposeful behavior. We will deal with this second item in Section 2.

The term 'decision' can properly be applied only to controlled, or cybernetic, systems, which are also designated as purposeful, or goal-oriented, systems. Although the category includes all living things and a variety of man-made servomechanisms, in this paper we are interested only in (1) individual human beings when they make choices in relatively complex matters and (2) formal organizations in which identifiable persons make choices that determine the coordinated behavior of the organization as a unit. 'Unit' for this purpose might mean, for example, either a whole corporation or government, or some formally designated sub-division of it, like a department.

By contrast, some other social units are uncontrolled systems. Their processes are essentially ecological, as with a market economy. Although interactions of individuals and sub-organizations within it produce a joint *outcome*, that outcome is no more a *decision* than the ecological balance in a forest is 'decided' by the oaks, bees, and squirrels.

Certain loose usage of terms occurs, and I would suggest usage as follows. When I make up my mind what to do about certain inadequacies of my rural

Hooker, Leach, and McClennen (eds.), Foundations and Applications of Decision Theory, Vol. II, 69–89.

water system, or when the owner-manager of a small business makes up his mind whether to add a new product to his line, a *decision* is called for, and decision theory is potentially applicable. At the opposite pole are the social *outcomes* of multiple individual actions in which each separate action might be subject to decisional analysis but the whole is not. There is no meaningful sense in which anyone 'decided' to tip a boat when all rushed to the same side, to pollute Los Angeles air by driving, or to bring on the Great Depression in 1929. And although some individuals or organizations might have deliberately tried to influence the Kansas birth rate or Labor Day accident rate in 1975, the rates that actually prevailed were not the result of a conscious weighing of costs and benefits by anyone. I think we muddy the language to call something a collective 'decision' when many, sometimes all, of the people whose individual acts are sub-units of a collective result neither know nor care that they are contributing to that result. In fact, it seems most unlikely that even one accident was 'decided', and through much of human history births were not 'decided' even within individual families. In part analogue to the way a good scientific statement should be falsifiable, we might suggest that some phenomena should not be referred to as 'decided' unless some identifiable decision maker(s) could be said *ex post* to have made an error – as that a ten per cent higher rate or a three per cent lower birth rate should have been chosen instead. Some identifiable entity that attempts to influence some variable might err in this sense. For example, a population council might conclude a decade later that it should have promoted one child per family instead of one-and-a-half. But that is very different from saying that 'the society decided upon' the 2.2 per cent rate that actually prevailed. Nobody decided it, and decision theory is not applicable to it.

Economic theory reflects this distinction clearly. It applies decision theory to individual firms and consumers, but shifts to supply and demand analysis to explain the behavior of a whole market. In fact, it is central to the theory of perfectly competitive markets that a market is an uncontrolled system and that a *decision* by any one economic unit is powerless to affect the whole-system *outcome*. To return to the broader problem, confusion arises when any social state or condition that actually prevails, or any social behavior that is actually executed, is thought of as a social decision or a social choice. I urge that we view such things as social outcomes or results but not as social choices or decisions. Outcomes alone, not decisions, can happen in uncontrolled systems or informal organizations. (A category of semiformal organizations is useful for other purposes but is not necessary here.)

When a specific conscious conclusion is to be reached in a formal organization by two or more persons it is customary to refer to it as a decision, and that usage will be followed here. Each individual weighs costs and benefits within himself and can reach his own decision about what is best. Depending on the organization there may also be some expectation that the things being weighed are the costs and benefits *to the organization*, not to the individual, and that the joint decision somehow transcends individual values. We will not assess here whether or to what degree that transcendence occurs, but will merely note that the end product is in important respects both an outcome and a decision, often with a tilt toward the former. Even though we call it a decision, to the extent it is an outcome in the sense identified above, decision theory is not applicable and other approaches must be invoked. Sections 3 and 4 will amplify and attempt to deal with them.

2. Intrapersonal decisions

'Intrapersonal' here refers to a decision made by one person or by some multiplicity of persons acting consciously and as a coordinated unit. However, as noted, the next section will deal more specifically with the question whether multiple persons can ever make a true decision in the sense dealt with in decision theory. To avoid those complications, for this section we will speak as if the decision maker is one person.

For reasons spelled out elsewhere (Kuhn, 1974, pp. 40–43) adaptive behavior (of which decisions are a subset) by a controlled system involves the following three ingredients, which are the logically irreduceable elements of adaptive behavior. First, the system must have some *information* about the state of the environment in which it exists. For systems with the level of information-processing capacity of the human, this includes information about alternative courses of action and the probable consequences of each. Second, the system must have goals or *preferences* as among its own possible future states; not all such states are equally acceptable to it. Third, the system must engage in *actions* which, within the state of its information, seem likely to lead to the more preferred rather than the less preferred states. 'Action', for this purpose, also includes inaction. Information, preferences, and actions are considered the logically irreduceable ingredients because one cannot imagine adaptive behavior which omitted any of the three. Systems language uses the names of detector, selector, and effector for these three functions of a controlled system.

These three functions or components constitute the structure of the behavior-selecting portion of a controlled system, and can be thought of as a framework for decision theory as follows. A decision directly involves the first two items, in which one's information about alternatives (perceived opportunity function, or detector) is compared with his goals (preference function, or selector). The third function, or effector, may or may not come into decision theory itself, which often simply assumes that the course of behavior selected will, in fact, be carried out. The relation of these three functions to decision processes is further elaborated implicitly in connection with their use regarding 'causal' explanations of motivated behavior, to which we now move.

'Cause' has been viewed in natural science as the conditions antecedent to a 'result'. However, 'cause' has long been a muddy concept when applied to volitional actions, as in the well-worn example, 'What causes my arms to rise?' Cybernetics offers a sensible format for this problem. The causes of a volitional act lie in the detector and selector states antecedent to that act. Within the cybernetic model this relationship can be stated in syllogistic form:

> *If* John contemplates the alternatives of raising his arms or leaving them at his sides; and
> *If* John prefers (all relevant costs and benefits being considered) to have his arms raised;
> *Then* John will raise his arms.

The problem has often been fuzzy because it was formulated as if reference to a future state, rather than to antecedent conditions alone, was a necessary part of the explanation. By contrast, the cybernetic model identifies more explicitly the information and preference states which are able to 'cause' volitional actions by virtue of their both existing in the system prior to the willed act.

In the case of conscious, volitional behavior the human system acts on the basis of its images of possible future states (detector) and of preferences as among those images (selector). Since the detector and selector states are both antecedent to the action the question then ceases to be the logically tacky one of future-as-cause and is seen instead as a scientific question in neurology and psychology – of how images and preferences together can cause actions.

Cybernetics helps here because its coverage includes simple goal-oriented mechanisms whose detector and selector states we understand, because we designed them. The household thermostat has a detector mechanism that registers the temperature of its environment, a selector mechanism that 'prefers' one temperature to another, and an effector in the form of a heating and/or cooling unit that alters any deviant temperature to or toward the 'preferred' state. An automatic regulator in a steel rolling mill has a detector that registers thickness, a selector that 'prefers' to keep the steel within certain dimensional limits, and an effector that readjusts the rollers when the dimensions stray beyond those limits. Unlike the systems studied in physics, chemistry, astronomy, geology, meteorology and any other natural science except biology, the two cybernetic systems just cited are *not indifferent* as among alternative states of themselves or their environment. Indifference is the distinguishing trait of the natural systems to which we apply traditional causal explanations, and its absence in cybernetic systems properly justifies our calling such systems 'goal oriented'. Those who wish can also insist that the thermostat or thickness regulator is future-oriented, and the essential reasonableness of such wording when applied to purely mechanical devices is the basis for concluding that cybernetics provides a more insightful approach to causal explanations of purposeful behavior. (Among the works on system analysis that could be used for background on this topic, Ackoff and Emery, 1972 is perhaps the most useful.)

Thus a causal explanation of a cybernetic system's behavior is not encompassed till we have identified its detector, selector, and effector states and mechanisms, especially the first two. Once we recognize the basic logic of such explanation, the same principle can be applied to cybernetic systems with more complex and less easily identified goal structures and information-processing mechanisms, like human beings. Then the explanation of why a cybernetic system behaves as it does is found by filling in the first two premises of the syllogism – that is, by specifying its detector and selector states in such a way that the ensuing effector action logically follows as a deduced conclusion.

Two of the more obvious reservations one might have about the above can be dealt with as follows. First, the model should be tied to the assumed actual content of detector and selector, if and when actual content can be distinguished from verbalized or fancied content. Second, the model should construe as relevant only their content at the moment the decision is made, not (for example) their content when a New Year's resolution was adopted six months earlier.

The cybernetic model also helps clarify the meaning of both behavioral and decisional approaches. Though identified more by examples than by definitions, the term 'behavioral science' is now widely used. As I see it, I would define behavioral science as the science of controlled, or cybernetic, systems, and a behavioral explanation as one that seeks to identify the content of the detector, selector, and possibly the effector subsystems of those systems under investigation. As noted, behavior is not satisfactorily explained until the contents of the three sub-systems show a valid syllogistic relationship.

Decision theory deals with the same general area, but in a different mode. Whereas a behavioral explanation typically seeks to learn the actual content of the detector, selector, and effector sub-systems of some specified real system, decision theory deals with the logic of the relationships among the sub-systems and with various techniques for improving the information acquired by the detector or for more carefully identifying the goals or preferences of the selector. The decision-theory techniques for improving the content of the detector currently include such items as linear programming, marginal analysis, computer scanning of alternatives, and game theory.

Behavioral science can deal with sources of preferences, with reasons why people perceive their available opportunities the way they do, and with the many possible interactions among the preference and perceived opportunity functions in a given system. Decision theory, by contrast, must take those states as given, and deal only with their consequences. It can only deal with such questions as: *If* (following due search and computation with sophisticated decisional techniques) your perceived opportunity function is so and so; and *if* (following due soul searching) your preference function is such and such; *then* your best alternative is this or that. In this respect decision theory is analogous to economic theory, which deals with the consequences of, say, technical coefficients (an opportunity function) and consumer desires (a preference function), but not with their origins. Without that limitation the theories of both economics and decisions would become hopelessly complex.

This limitation of decision theory is mentioned because, for reasons amplified below, decision theory must be supplemented by social and behavioral science before decision making at the individual level can be linked with decision making at the social level – that is, with social choice. Economics avoids this difficulty in theory by deriving product supply and demand curves directly from preference functions of firms and consumers, and by next deriving equilibrium market prices and quantities directly from supply and demand curves – though I consider economic theory remiss in giving so little attention to the way consumers actually make decisions, particularly as influenced by advertising and other merchandising activities.

Decision theory is at its best in connection with maximizing solutions. While theoretically neat, I doubt that the maximizing models fit many real decisions that are relevant to social (or even individual) choice. A value can be maximized only within constraints. When the constraints are external and cannot be altered by the decision maker – as with the law, the competitive price of wheat, gravity, or the 24 hours in a day – no problem arises about the notion of maximizing within constraints. Often, however, constraints are themselves within the control of the decision maker and consist of alternative goals. In that case the problem takes on a different logical quality, which to me seems better handled as one of satisficing. For example, while recognizing that each of the three ingredients is subject to its own external constraints, it makes no logical difference whether I ask: 'What is the most fun I can have without unduly jeopardizing health and reputation?' 'What is the best reputation I can have without unduly jeopardizing health and fun?' or 'What is the best health I can have without unduly jeopardizing fun and reputation?' The problem is to achieve a satisfactory balance among the three.

I can, of course, formulate this question mathematically, in which U_f, U_h, and U_r are the utilities to be received from fun, health, and reputation, and U_t is a total. I can then set $U_t = U_f + U_h + U_r$ and say that the problem is to maximize U_t. By the same logic I assume that it is theoretically possible to reformulate any similar problem of satisficing into one of maximizing. The fallacy of all this – and it seems to me to be of a sort that recurrs often in this conference – is that just because we can write a theoretically solvable formula for something we tend to assume that real people can *do* it, and that we have said something meaningful about reality instead of just writing down some symbols. In this case I suggest that the kinds of satisfactions to be achieved from fun, health, and reputation are qualitatively different. The real decision maker juggles among the three goals until he finds a level of each that is satisfactory in light of the other two – and then he leaves it. More specifically, I think I recognize pieces of reality that correspond meaningfully to U_f, U_h, and U_r. But I do not think I recognize any piece that corresponds to U_t as here defined. (However, I do recognize U_x, a special and independent joyous satisfaction from finding any solution at all that avoids a really unsatisfactory level of any one of the other three.) Hence it strikes me that when we reformulate this problem of satisficing among three separate utilities into one of maximizing a composite total utility we move from a real problem to a mythical one. Whether apples and oranges can be added when we subsume both under 'fruit' cannot be answered in general theory, but only with reference to a given real problem. For *this* real problem I do not think they can be added, and hence urge that we continue to view the problem as one of satisficing. Other cases may be different.

According to the economist one can maximize total satisfaction from a given total of money expenditures by equating the marginal utilities of all expenditures across all alternatives. For real people actually to equate at the margin (in contrast to the economist's theorizing about it) is difficult enough with expenditures of money. In most of life's decisions the costs as well as the benefits are qualitatively different for each alternative (How much sunburn am I willing to accept and how much reading am I willing to forgo for another hour of swimming?). Here, too, we must avoid thinking that something can be accomplished in reality, even approximately, just because we can write a formula for it.

Another hurdle to maximizing is that the decision process itself has costs, in time, effort, and frustration, and the improvement in result may simply not be worth the cost differential between maximizing and satisficing. These costs can be social as well as psychological, as friends impatiently say, 'Make up your mind!' Even in supposedly objective decisions, as in selecting a corporation's pricing policy, things may not be so clear. Let us take the following situation.

The decision maker already knows that Alternative *A* will be satisfactory, and to accept it without further ado constitutes a 'satisficing' solution. Omniscience knows that Alternative *B* would be far better. But the decision maker does not know even roughly either the probability that *B* will be better or the amount of its superiority. Nor can he know till after he expends it whether the cost of the information needed to know either of those things will exceed its benefit. If 'maximizing solution' is taken literally to mean the set of activities that will lead to the selection of *B*, then we can certainly accept that 'To optimize requires processes several orders of magnitude more complex than those required to satisfice' (March and Simon, 1958, p. 141). However, if 'maximizing solution' is taken more flexibly to mean that we should toss a coin under this kind of uncertainty, we cannot accept even that conclusion until we first put some valuation on risk. If the decision maker places a high value on risk avoidance even the maximizer will choose the 'satisficing' solution, and select it over the 'maximizing' solution of tossing a coin. This is only one of many reasons why I would urge decreased attention to maximizing, since I doubt that is relevant to very many of life's real decisions. Relatedly, before we assess the merit of collective decisions as aggregators of individual decisions we need more empirical information about the nearness-to-optimum of the individual decisions being aggregated. To use a mathematical analogy, we must not try to compute a five-significant-figure average of two-figure data.

Relatedly, decision theory is useful mainly for fine tuning. If the total area of possible alternatives is represented by a 26X26 matrix, with rows *A-Z* and columns 1-26, intuition, global impulse, normative constraint, or some other relatively vague criterion tells us that the choice lies somewhere around *K*-6, not near *A*-2 or *Y*-22. Formal decision techniques can help ascertain more precisely where in or near *K*-6 the optimum point lies; but they are of little help in bringing us to the general neighborhood of *K*-6 to begin with. For example, if a business competitor has become troublesome, formal decision theory does little to identify that the major alternatives are to cut prices, sabotage the competitor's factory, or seduce his wife. Yet the most important step in any decision is likely to be that of identifying the approximate locale of the possible answers.

One further limitation of decision theory is relevant to the present topic. Both one's preferences and his perceptions of the available opportunities often change as a consequence of thinking about making a decision, as a consequence of thinking about effectuating it, or both. The actual process may thus be dynamic, not static, and the changes may affect the decision maker or his environment. When or whether to stop mid-passage to reassess the decision situation in light of such changes is itself a major part of the decision. But because such changes depend on the person, the situation, and chance, decision theory can tell us little about them.

To summarize this section, decisions in the sense dealt with by decision theory occur only in controlled, or cybernetic, systems. When related to a system-analytic view of the structure of a controlled system we can say that a causal, a behavioral, and a decisional explanation of the system's behavior are basically the same, even if with differences in focus and coverage. Recognition of this relationship should help us link various kinds of knowledge toward an understanding of social decisions.

The potential of decision theory for improving the quality of numerous actual decisions is real, and need not be detailed here. To provide a link between decisions at the unit level and those at the social level nevertheless requires us to recognize several limitations of decision theory: that it (like economics) must take preferences as given, that it is not at its best in dealing with the satisficing decisions that are of most concern to social decisions, that it is more useful for the fine tuning than for the important global estimating of solutions, and that it is not much help in cases where the important variables undergo continuous dynamic change while decision making is actually in progress. For such reasons we need to join decisional and behavioral theory in a supplemental relation if we seek a successful link between the individual and the social levels of making choices.

3. Small Group Consensus

This section deals with the achievement among a small group of some consciously formulated decision(s) that will be binding on all members, in the face of some initially incompatible desires among members about means, ends, or both. 'Small group' here means that each member can observe the positions taken by and can potentially interact with each other member. An understanding of consensus processes within the small group is a logical link between decision theory at the intrapersonal level and any realistic understanding of larger scale social choice.

Consensus can be reached in such a group by any one, combination, or sequence of three processes: communication, transaction, or dominant coalition. (See Kuhn, 1974, pp. 306–313, based on a variety of other analyses and categorizations of these processes.) To illustrate the first, in pure communication the members talk, write, argue, or otherwise transfer messages as a means of changing the preference or perceived opportunity functions of other members. If completely successful the result is unanimous agreement that one particular course of action is best – the non-market analogue of Pareto optimality. This means not merely that it is accepted by some as the best they think they can get agreement on, but that it is the most preferred alternative of every member.

To the extent the members do not all come to desire the same outcome – to the extent consensus is not achieved by communication – they may nevertheless reach agreement by transaction, which means exchange, tradeoff, or compromise of values, and may include side payments. As contrasted to consensus by communication, in which everyone gets what he wants most, here it is possible that no one gets it. But it is still possible for each member to agree to the solution if he receives Pareto-improving concessions adequate to compensate for the ones he gives. Though no one may be entirely happy, and some may feel they got a bad bargain, the solution may still be accepted by all as reasonable under the circumstances. Because consensus by transaction requires concessions, whereas concensus by communication does not, the latter is presumably the preferred technique if it can be achieved. Other things equal, it will allow a larger total utility.

Suppose that both processes run their course without producing agreement. This means that at least some members can neither get what they want nor get concessions that seem adequate to compensate for not getting it. Under these circumstances *agreement* is not induced. However, that does not mean that a *decision* is not possible, and the remaining technique for achieving it is the dominant coalition. The best known form is the majority vote, but depending on the circumstances the decision may go to that coalition (including the possible linguistic anomoly of a dominant coalition of one) that can outfight, outrun, outspend, outdrink, out-talk, or simply stay awake longer than any other coalition. The decision is imposed by the winners on the losers without their consent, though consent of sorts might lie in a prior agreement that any decision reached by an agreed-upon technique will be considered 'accepted'. Dominant coalition is presumably less satisfactory overall than the other two techniques. For while the winners win, the losers neither get what they want nor receive adequate compensating concessions. Whether or not this solution is 'dictatorial', it is 'imposed'.

The possibilities within and among coalitions are vast, and the study of them might well be considered a discipline of its own. We will mention only two of its more obvious areas of attention. First, whether dominant or not a coalition is likely to contain subcoalitions. For example, in a group of twenty with a dominant coalition of twelve, the chances are large that if two or three leaders of the twelve take a firm position the remaining nine or ten will go along. Larger groups are likely to have additional hierarchical levels, and actual control of the whole group's decision may lie with those few who dominate a sub-sub-subcoalition, especially if they arrange cross ties with those who dominate other coalitions at various levels.

The second area of attention concerns the conditions under which a person or a subcoalition may switch coalitions. For example, when does one switch from a losing coalition that he dominates to a winning one that he does not dominate?

With this background let us now return to a central question in social choice – namely, the degree to which a group decision reflects or represents the preferences or values of the individuals in the group. In consensus by communication the answer is clear – everyone gets what he wants. In consensus by transaction the answer is uncertain. If the initial divergences were small and the concessions minor nearly everyone may be pleased. If the divergences and concessions were large nearly everyone may be displeased, relative to his preferred position, but accept the outcome as the best available. It nevertheless remains true that the transacted solution can in some way or degree embody and reflect the preferences of each group member, though a narrative history of the negotiations might be necessary to trace the relationship.

By contrast, almost nothing can be said about the relation between the members' preferences and a decision by dominant coalition. For example, to what extent did individuals already compromise within their respective coalitions? If the main decision is by vote, what tradeoffs occurred before the issue reached the form that was voted on? Are the members of the non-dominant sub-coalition within the winning main coalition better off than if they had switched to the dominant position in the losing coalition and thereby made it win? Are the losers only modestly worse off than if they had won, or is their loss disastrous? Is the loss perhaps modest to half of the losing coalition and disastrous to the other half? The questions are much the same if dominance is based on force, money, media coverage, or some other criterion instead of votes. Yet these are only a few of the questions we need to answer before we can say much about the group decision as a reflection of the preferences of its members.

To illustrate a further complication let us look at a group discussing a possible decision. Jones mentions a solution that he considers potentially useful. But he drops further advocacy of it when he receives what he thinks is a *sub rosa* cue from Smith to change the subject, because he respects Smith's judgment. Windsor and Stuart also consider Jones's suggestion meritorious, and are ready to support it. But they respect *his* judgment, and assume that Jones abandoned his proposal because further thought had convinced him it was not workable. The group then reach a decision by ballot without including the Jones proposal among the alternatives. Whether or not Smith intended to silence Jones, the relation between the group decision and its members' preferences is, to say the least, murky.

Even in formal organizations where authority flows down and identifiable individuals are supposedly responsible for identifiable decisions it can often be remarkably difficult to find out who actually *made* a given decision. We are not speaking here merely of attempts to hide blame, but of cases where conscientious participants are genuinely puzzled to understand how some course of action was actually selected. Even in carefully structured situations decisions are often *outcomes* that *emerge* rather than being *made*.

The small group is a useful vehicle for learning about social decisions. Its size permits both the participants and outside observers to encompass all overt aspects of the process. It also permits the observers by questioning or projective techniques to learn much about what goes on inside the participating individuals. It is further amenable to controlled experiment in which selection of individuals, formulation of the problem, rehearsed stances taken by certain individuals in the group, and other items are determined in advance by the experimenter. It is abundantly clear that, whether or not the decision process culminates in a vote, the social processes and structures have a great deal to do with the ensuing decision and with its relation to the initial value structures of the members. In the small group it is also possible to poll the sentiment in advance of any interactions and thus to get a direct reading of initial preferences. By contrast, there is rarely any point at all to taking a poll of the general public till after there has already been substantial interaction about an issue, and little weight would be assigned to the poll until such interaction had occurred. Hence the conclusions of this Section 3 become stronger, not weaker, as we move from the small group to the larger public.

It is abundantly clear from the small group that to understand how we get from individual preferences to a group decision requires not only general social and behavioral science knowledge but also knowledge of the particular situation. There is the additional sobering question of whether we can ever get far beyond saying that the public will *is* whatever does, in fact, emerge from the extant institutions for making decisions on the public's behalf, as one sees from a reading of Rothenberg (1961, p. 318). It is, of course, conceivable that a group might first reach a consensus about the costs of each alternative, a separate consensus about benefits, and then select among the alternatives as a third step. Whether or not that technique is feasible for certain limited-scope decisions in relatively small collectivities, it is a situation hardly ever faced for a whole society. In most techniques actually used *outcome* seems a more congenial term for the end result than does *decision.*

4. The Larger Social Choice

The purposes of this section are (1) to identify more explicitly but briefly the relation between the conclusions of the preceding section and the conclusions of Kenneth Arrow about social choice, (2) by example to show how the massiveness and complexity of a pre-existing situation place grave limitations on the scope of social alternatives about which a voter can even be said to *have* meaningful preferences, and (3) to identify briefly the small information-carrying capacity of the ballot compared to what a voter needs. Together these limitations seem to me to dwarf the significance of those delineated by Arrow.

In connection with the 'paradox of voting' and related items Arrow has rendered an important service in demonstrating grave reservations about the ballot as a 'social choice' which meaningfully represents the values of the individual voters (Arrow, 1951). That his conclusion is based on the assumption that we ignore intensities of preferences is probably sensible, in light of the unsophistication of many voters and the cost of counting ballots. More recently he has incorporated additional items into that relationship, while also going into other problems (Arrow, 1974).

We need to understand the limited meanings of ballots, and Arrow has stated them elegantly. Hence it is perhaps proper that 'impossibility theorem' and 'paradox of voting' have been mentioned so often at this conference. But for reasons already stated and amplified below I think the time now overdue to recognize that whether or not a ballot accurately reflects voters' values it is still only a small part of the total process of translating individual values into social choices – typically a tiny point at the apex of a pyramid of decision steps. Perhaps better, there is a large family of possible decision pyramids. For most potential ones nothing more than the base is ever built, if that. Others get farther along before they are abandoned. But only an occasional few get far enough to add the ballot at the apex. Meanwhile others are completed by techniques that do not include a ballot at the top, and that may or may not include voting among the sub-processes. Pieces of some decision pyramids are transferred in huge chunks to other pyramids, and others sprawl indiscriminately over or between indeterminate additional ones. To say that a vote, when taken, shows little about the underlying preferences of the voters is true, but a distinctly minor part of the problem.

In discussing the public will there is often an implicit assumption that a voter, at least if he is intelligent and well informed, holds a set of mental images about alternative solutions to a given problem, and at least a rough rank-ordering of his preferences among them. While such orderings may be developed by students or professors for academic purposes a voter rarely generates one or is able to make all the required pairwise comparisons. Typically instead, the voter faces a situation so laced with complexities and givens that only insignificant details are open to be influenced by his vote. We will use local street transportation to illustrate.

Some people drive automobiles, on streets or highways financed partly by taxes on gasoline, partly by general taxation, and partly by direct toll charges. Various segments of the taxes have been decided by legislative bodies at local, state, and federal levels, some recently and some perhaps two decades ago. The decision to build a particular section of highway has been made by local, state, and federal departments of highways or transportation, possibly modified by an environmental impact study, by a long-range regional planning commission, by a referendum in one or more suburban municipalities through which the highway runs, or by kickbacks from a highway contractor to a county commissioner. A given person's decision whether to drive or go by bus is influenced by the price of gasoline (which also depends nowadays in several ways on decisions made by the president, the Congress, and by state legislatures), by the cost of parking (which may be done in private or public lots and which may or may not be subsidized by taxes, or on the street in spots which may or may not be metered at prices set by the town council), and by the cost of riding a bus (which may also be subsidized by federal, state, or local governments through decisions made by legislators, by administrators, or by referendum). Perhaps another alternative is for the person to ride the commuter rail line of a semi-public corporation which is subsidized at the federal and state levels and which leases track and equipment from a bankrupt railroad which itself lives only by courtesy of federal loans and some leniency in applying the bankruptcy laws. The cost of those tracks is also borne in part by freight shipments over the same rails at prices approved by a federal regulating commission.

Under these conditions voters learn that a candidate for city council strongly favors municipal ownership of the local bus company. The same candidate also opposes busing for racial integration and supports using welfare funds for abortions. When citizens contemplate voting for or against this candidate, in what sense can their decisions be said to represent their individual preference orderings as among all possible techniques for local passenger transportation? The question is almost ridiculously irrelevant.

How much better are voter preferences on this matter reflected if they vote directly on the issue? Presumably somewhat better than if they vote for a candidate. At least it is clear what issue they are responding to. But referendum issues are not self-formulating. Some persons or processes must first determine whether a matter is to be put to a vote at all. Persons or processes must also determine how the issue is to be formulated for the vote: that it will be presented as a yes-or-no vote on municipalization of the local transit company instead of as a vote on subsidizing the private company, encouraging the growth of satellite communities to reduce downtown traffic, or constructing more arterial freeways. These matters are determined (in the present model) through a network of interrelated communicational, transactional, and coalitional interactions among a variety of persons and groups at a variety of levels. The point is that before the issue even gets onto the ballot all the most important parts of the social decision are already made and foreclosed. The ballot gives the appearance of importance because it is the visible peak that completes the decision pyramid. But if we are seriously interested in the relation between individual preferences and the actual course of social action we should devote most of our attention to the large bulk that lies below the apex.

The problem of focus is not very different for candidates then for issues, since all the same questions can be raised about the lower level decisions: about who should be considered as possible nominees, which ones the party will support with how much money and enthusiasm, which issues will be played up or down, which blocs of voters will be appealed to, and so on. All these things must be known if we are to interpret the meaning of an election, and the paradox of voting barely introduces us to the problem.

Information theory provides another context for this question. A voter communicates one 'bit' of information when he votes yes or no on an issue or chooses between two candidates. He would communicate to bits of information if he selected among four alternatives, but that is relatively rare. In the case of a referendum, if he has thought much about the issue he would probably need at least twenty bits in a well-constructed listing of alternatives to identify his real preference. Relatedly, I have voted many times when my first, second, and third preferences were not even listed on the ballot, and wondered what I was communicating. Why, then, do we pay so much attention to the particular weaknesses of one- or two-bit ballots, which are the kind on which the Arrow theorems concentrate? The real concerns about the relation between individual values and social choices lie mostly elsewhere.

5. Pareto and Optimality

I was much impressed with how often Paretian optimality was mentioned at this conference, as if the term has some meaning or facilitates some understanding. Here I think the time long overdue to assert that the notion has very limited usefulness, if any, in connection with real problems of social choice and resource allocation. Let me deal first with seeming exceptions.

First, it is useful to know that almost any society has room for many Pareto-improving trades, and that total utility will be raised if they are made. But that conclusion also follows from an understanding of the simple selfish transaction, whether or not the notion of Pareto optimality for the whole collectivity had ever been formulated.

Second, the concept is useful as a way out of the logico-theoretic impasse created when we disallow the Hobbesian addition of utilities among persons, since it identifies an optimum that does not depend on comparing utilities interpersonally. Further, the Paretian concept is logical and clear when applied to two persons and two tangible commodities, as in the Edgeworth box, and may remain reasonably so among several persons and several commodities. However, it seems to me fallacious to assume that the idea remains qualitatively valid when the quantities of persons and commodities expand indefinitely. Let us observe several points.

In a two-person two-commodity world it is clear that a move to the Edgeworth contract curve from an adjacent point not on the curve will leave at least one party better off without worsening the position of the other, and may leave both better off. But if this concept is to relate in some way to reality, rather than being a mere logico-mathematical exercise, some mechanism must exist that is able to produce this result, which means also that the parties must be able to make a series of pairwise comparisons between sets of alternative positions. In the two-person two-commodity world I find such a mechanism wholly credible: each person need know only his own utility function and keep bargaining till the pair reach a mutually acceptable solution. While the abstract logic remains the same as we add many people and commodities the notion of a mechanism through which all pair-wise comparisons and bargains could be reached quite boggles the mind, much less the capacities of real space-time bounded persons. Granted that the notion of a perfectly competitive market is a great brain saver in this respect, it, too faces grave limitations. For one, the conditions for a perfectly competitive market are extraordinarily rare in reality, and even conditions that are reasonably close characterize only a small fraction of any real economy. Second and relatedly, many of life's utilities are not really marketable, because they are local and particular, and any trade-offs involving them would require explicit person-to-person negotiations, not mere separate responses to a market price. For example, externalities in the form of utilities or disutilities to passers-by or neighbors from my beautifying or not beautifying my property may be represented by a symbol in a formula. But unless we can conceive a mechanism that will measure and allow trade-offs among all such utilities on a society-wide basis (which I cannot conceive) I see no practical application for the formula.

Perhaps the concept of communication net will help clarify. A perfectly competitive market is a wheel network. The market price is at the hub, with participants around the rim. Each party can learn all he needs to know by communicating with the hub, each channel constituting one spoke. By contrast, successful effectuation of a Paretian optimum that also covers the many local and particularized goods would presumably require an all-channel network, in which each participant makes or contemplates trades of at least some commodities with every other participant. The theoretical beauty of the perfect market is that the number of communication channels rises at exactly the same rate as the number of participants, whereas channels in the all-channel network rise as the factorial of that number. The number of channels becomes unmanageably large relative to the task to be done when participants number more than five or six, and becomes utterly ridiculous when we contemplate societies of hundreds, much less of millions.

To generalize, and explicitly including communication channels as part of the mechanism, I suggest that before we expand any concept from two persons and dimensions to n persons and dimensions we first identify the mechanism by which the conceptualized result is actually to be implemented in the former case, and then demonstrate that the mechanism is no less expansible to the latter case than is the logic. I suspect that such a requirement would have shortened this conference considerably.

Let us now look at Pareto optimality more specifically as applied to any real, large society. This contrasts with the microcosmic details where it is easy enough to know that utility is increased when I swap some of my surplus tomatoes for my neighbor's surplus sweet corn. But these reallocations are trivial when we think about significant reallocations in a society of millions – as among such blocs as workers, farmers, small business, Southeast or Northwest, the elderly, students, and minorities, and lead to several observations.

First, for many reallocations made through conscious policy decisions, as with welfare payments to the poor or relief to the distressed, the policy is more Hobbesian than Paretian. That is, we feel quite sure that utilities for the redistributed items are greater for the recipients than for the donors, and we do not need Pareto. Second, if not made on this basis, the reallocations are probably reflections of political power – another non-Paretian criterion.

Third, when we move from the ivory tower notion of a theoretical optimum to any proposal for significant redistribution that is likely to be suggested or implemented in the real world we find that such proposals nearly always have the explicit intent of making some group(s) worse off as the means of making others better off. But if so, then so far as we can tell the *status quo* already *is* Pareto optimal relative to the outcome of the proposed change. That conclusion can, I think, be generalized to virtually any realistically probable proposal about reallocation. If so, we face the nasty conclusion that for practical purposes any condition actually faced by any society must be treated as if it already is Pareto optimal relative to any proposed major change. Considered as a guide to real policy making I thus find it hard to imagine a more useless concept. In addition, any such optimum is local. As Boulding (1952) once noted, it tells us nothing about when or whether to climb down a molehill so we can start climbing a mountain.

As an additional difficulty, the Pareto concept is logically clearest as regards allocating a given total of tangible goods. If we can add to the total with no increase in or reallocation of costs we can be sure of raising total utility. But (if we take into consideration the question of measurement discussed shortly below) Pareto optimality gives no clue as to where to allocate an increase of goods so as to produce the *largest* increase in social welfare, since *any* allocation will make the recipients of the increase better off without making anyone else worse off. The notion virtually dissolves when we also consider non-marketable intangible goods and public goods and bads. For example, some segments of the population want the intangible and largely non-marketable utilities of security of their properties and of their persons on the streets. Other segments want the intangible and non-marketable utility of the fun of vandalizing and molesting. Under the Paretian criterion, how are we to know whether we have moved toward or away from the optimum if we provide the first set of utilities at the cost of denying the second, since we have obviously made one group worse off in the process? How do constraints on air pollution affect total worldwide welfare if their marginal real per capita cost is very high in poor nations and very small in rich ones? (Which raises the incidental question whether Pareto optimality is nationalistic.)

Finally, while it provides an elegant abstract logical escape from interpersonal utility, and may be a guide to some marginal improvements of detail, Paretian optimality, as such, is absurdly useless for real social problems, since there is no possible way of measuring it. For example, how would one proceed to determine whether the United States is or is not now Pareto optimal? Economic statistics are of no help, since they measure public observations, while the optimum is concerned with utility, which is private. True, we might conclude that allocations in some industries might be closer to optimal if the industry were more competitive. But one cannot change the structure of an industry without having possible effects on its productive efficiency, rate of innovation, and other variables, so there is no guarantee that the change would actually be an improvement. Should data on crime, divorce, and automobile accidents be considered? Would a questionnaire provide relevant data? And if so, what questions would we ask, of what sample of the population, and how would we interpret their answers? Among other things, should people's *estimate* of the effect on their level of satisfaction of some change in allocation be taken as a satisfactory proxy for the effect that would actually occur? Among all possible reallocations, which should we ask about, and why? These examples only start the problems of measurement. For such reasons I see no conceivable way of knowing whether we are or are not at the optimum, or of the direction or magnitude of the deviation if we are not – that is, of making any practical use of the concept.

For the above reasons, among possible others, the meaning of Pareto optimality resembles that of unicorn, phlogiston, and ether. While we may have a mental image that is evoked by the term, it seems doubtful that there is any real state of any whole society that we can identify as an example of it, at least above the level of a few persons and a few goods. Except for its possible limited use as a heuristic or pedagogical device, or for making small changes that could probably be handled without it, I think we should seriously ask whether the Pareto optimum should be buried in phlogiston and covered with ether.

6. Summary

Considering the small part that voting plays in converting individual values into social choices, when viewed in wider perspective perhaps the real paradox of voting is why the paradox in Arrow's sense commands so much attention. Its importance, though great, is now mainly historical. By a demonstration that the ballot box often will tell us little about the 'will of the people' – or indeed, that often there is no such will – Arrow helped break down a social and psychological barrier to a more realistic examination of social decision processes. Nothing in such observations reduces the immense practical need for the sobering effect on politicians of knowing that there is always a next election. In fact, *not* knowing what will get the voters' backs up may be more sobering than knowing.

Second, decision theory must be heavily supplemented by social and behavioral theory if we are to trace any significant linkage between preferences of individuals and the actions actually taken by their society. Specifically, individual preferences become social actions through the communicational (including cultural and normative), transactional, and coalitional structures and processes of the society, and a ballot is only one of many mechanisms of coalition. These structures and processes act and interact with one another, among many subparts and in many sequences. To mention a conspicuous case, it would be ridiculous to contemplate the overall relation between individual values and social choice without careful attention to the power structure of the society. These processes are always dynamic, with the decision processes changing the preferences and vice versa. As a microcosm version of social decision making the small group displays the tenuous relation between individual preferences and a collective outcome. The tenuosity rises rapidly, probably exponentially, with the size of the collectivity. This is not to suggest that only social and behavioral scientists should deal with the problem, but merely that to ignore the social and behavioral components is virtually to ignore the problem.

Finally, it is fallacious to generalize (as we so often do) from a two-party two-dimensional relationship to *n* persons and dimensions unless the mechanism (usually implicit but sensible) that operates in the simpler case is identified and is demonstrated to be operable in the complex case as well. This fallacy, along with other difficulties, suggests that the concept of Pareto optimality be thrown out as applied to any complex society, to be reinstated only with great circumspection, only for very limited applications and only by those who understand why it should be thrown out.

University of Cincinnati

BIBLIOGRAPHY

Ackoff, R. L. and Emery, E., *On Purposeful Systems,* Aldine-Atherton, Chicago, 1972.

Arrow, K. J., *Social Choice and Individual Values,* John Wiley and Sons, New York, 1951.

Arrow, K. J., *The Limits of Organization,* W. W. Norton and Company, Inc., New York, 1974.

Boulding, K. E., 'Welfare Economics', in B. J. Haley (ed.), *Survey of Contemporary Economics,* Richard D. Irwin, Inc., Homewood, 1952.

Kuhn, A., *The Study of Society,* Richard D. Irwin and the Dorsey Press, Homewood, 1963.

Kuhn, A., *The Logic of Social Systems,* Jossey-Bass Publishers, Inc., San Francisco, 1974.

Kuhn, A., *Unified Social Science,* The Dorsey Press, Homewood, 1975.

March, J. G. and Simon, H. A., *Organizations,* John Wiley and Sons, Inc., New York, 1958.

Rothenberg, J. *The Measurement of Social Welfare,* Prentice-Hall, Inc., Englewood Cliffs, 1961.

R. D. ROSENKRANTZ

DISTRIBUTIVE JUSTICE

1. Introduction

Theories of distributive justice, as I conceive them, are primarily attempts to answer the normative question: what division of burdens and benefits attaching to a cooperative endeavor is fair? I wish to consider a theory that proportions distributive shares to marginal product, making due allowance for the 'costs' to the contributor. By *marginal product*, I shall mean the value of what the contributor adds to what is being produced without him. The present formulation is certainly vague in the absence of any indication what allowances are 'due' the contributor for the costs he incurs. I will only advance to a more adequate statement of the theory by stages. It will be well to begin by clearing the ground of those misconceptions that are most likely to arise at the outset.

One of the suspicions that confronts any theory of justice that departs from what one may call the 'market theory of value' is that fair distributions will be *imposed*. They need not, however, be independent of what people actually desire. I have already defined 'marginal product' as the *value* one's product adds to what others are producing, and I certainly intend that to mean the value others attach to one's product. One cannot assume, however, without begging the question, that the value of product is equal to its *market* value. The rich may pay more for given luxuries than the poor pay for given necessities, but it hardly follows that the rich value their luxuries more than the poor value their necessaries. The real problem here, of course, is to imagine alternatives to the market mechanism for distributing burdens and benefits, or setting prices and wages. One of my purposes is to suggest such alternatives.

Similar remarks apply to 'costs'. Under this rubric, one might include working hours, the required amount of prior training for the job, health hazards, and the like. Even from this short list, the difficulty of objectively assessing costs must seem formidable. Are we, for example, to reward a man for his effort? That is, should society pay a less skilled person more to perform a task merely because it costs him more effort to equal another's output or match the quality of his produce?[1]

Fortunately, these difficulties can be swept aside. In the last analysis,

Hooker, Leach, and McClennen (eds.), Foundations and Applications of Decision Theory, Vol. II. 91–119.

there will be no need to pin down the meaning of 'cost' (any more than the meaning of 'value'). All one need mean by 'cost' is the relative disutility a man attaches to specified work or loss of leisure, as crudely measured, perhaps, by the amount he is willing to pay to do one job rather than another. People will obviously differ, then, in the disutility they attach to the same work. (Some abhor hard physical exertion; others relish it.) Hence, it needn't be the case, from this perspective, that the wage rate for given work be uniform over persons. I will return to this point.

2. Efficiency

Efficiency is usually put forward as the strongest argument for a free market. How else could labor be more efficiently allocated to satisfy consumer demand? Imagine a firm faced with the problem of maximizing net returns. It must decide what production factors and services are needed and at what price to purchase them. A familiar bit of reasoning shows that an optimal allocation sets wages or factor costs proportionate to marginal product over all items purchased. If the payment of an additional dollar to secure more of a given skill or service netted a higher return than the last dollar spent on another, the allocation would not be optimal and the firm could go on increasing its return by substituting more of another item until equilibrium was restored. Now if we think of society as a large firm in the market for services and goods, what follows is that labor should be remunerated in proportion to its marginal product.

It will be argued that an ideally free market tends to this end result. But the price the market attaches to a service is no simple function of marginal product. It depends, as well, on the availability of others who can provide that service. One might reasonably argue this is as it should be. For example, it costs more to train a doctor than a garbage collector or a bus driver, and hence the greater cost of replacing a doctor should be reflected in a higher incentive to keep him on the job. Nevertheless, once the relevance of such considerations is admitted, we have strayed from any strict adherence to marginal product.

From the standpoint of a consumer, efficiency is certainly a desideratum; we want the best qualified man for the job. When we assume the standpoint of a worker, however, we become aware of a sharp cleavage in our attitudes. If the qualifications of two men for a desirable position differ only slightly, it might seem more reasonable to share the position between them, if that is feasible. In any case, there is another sense of 'efficiency', a welfare sense, in which it is not very efficient to pay more for a service than the man performing it is willing to accept, rather than abandon it for another job.[2] People are sometimes heard to say that they feel positively immoral about being paid to do what they are doing out of love. At the other extreme, when men are called upon to labor in an outpost of civilization, they receive 'hardship pay'. One might argue that such inflated wages are justified from an economic standpoint as necessary incentives. One can take a different view, however, and regard them as given in recognition of the extra 'costs' born by those who labor in such conditions. But from this point of view, coal miners and garbage collectors are as entitled to 'hardship pay' as men who work the Alaskan or North Sea oil rigs. Those who think that coal miners or garbage collectors are overpaid, or that their salary levels are artificially high as the result of collective bargaining, should ask themselves how high a wage it would take to induce them to exchange their present job for one of those.

Suppose, indeed, that everyone had identical skills and tastes. That would not preclude strong biases against certain kinds of work. A fair scheme of wages, I claim, would leave everyone absolutely indifferent between doing any two of the jobs at their assigned wages.[3] Workers could then be assigned their jobs by a social lottery. It would be hard to quarrel with the fairness of such a lottery. Each worker would be able to justify his position to the others; indeed, each would be willing to exchange his position for any of the others. If that is so, it lends some support to my contention that, under a just distribution, socially dispreferred positions would receive higher remuneration. I think that this tendency would continue to characterize a just distribution when tastes and skills varied, but I will withhold my reasons for thinking this for the present. It is clear that the cited tendency rarely characterizes a market economy. The least desired jobs are often perversely saddled with the lowest wages. For the supply of labor for unskilled jobs tends to be higher than the demand, so that unskilled wages are bidded down. If that is less true today than it was at the height of the Industrial Revolution, the various interferences with a free market – collective bargaining, minimum wages, and so forth – are at least *arguably* the cause. That so many are still wont to regard these interferences with voluntary exchange or freedom of contract as unjust or unwarranted testifies to the strength of the tendency to tautologically identify 'value' with 'market value' and 'fair return' with 'market return'. It is conceded, at most, that a more egalitarian distribution of income and wealth is a desideratum to be weighed against the *conflicting* desiderata of efficiency and individual liberty.[4]

Now everyone would like to believe that *good things*, like justice, liberty, and efficiency, can in principle coexist. It is among the aims of this chapter to determine the extent to which this is really so, but the question cannot be fruitfully addressed until the implications of our theory of justice have been laid bare and its practical implementation discussed. As it happens, my views emerge rather naturally from a consideration of the rival contractarian theory of justice that has been developed by John Rawls. The discussion of this theory will occupy us in the next two sections.

3. The Difference Principle

John Rawls' early theory of justice (Rawls, 1958) admits inequalities only when they benefit everyone. This requirement is symmetrical with respect to the cooperating parties. It also disallows utility trade-offs of the kind (apparently) sanctioned by Benthamite utilitarianism and various compensation tests, namely, large gains to some which are allowed to offset or 'compensate' losses to others. Rawls' later formulation is non-symmetrical and admits inequalities only when they benefit the least advantaged. This is hedged somewhat by the assumption of chain-connectedness, which holds that benefits to the least favored percolate upwards and thus redound to the advantage of everyone. But where this empirical assumption does not obtain – and Rawls is quite explicit about this – 'the stricter interpretation of the difference principle should be followed, and all inequalities should be arranged for the advantage of the most unfortunate even if some inequalities are not to the advantage of those in middle positions' (Rawls, 1967, Section 4). Moreover, the advantages of the more fortunate most promote the well-being of the least fortunate in the strong sense that 'a decrease in their advantages would make the least fortunate even worse off than they are' (*ibid.*).

Rawls appears to have been led to his conception by considerations of reciprocity and desert. He argues that the distribution of good things should not depend on social contingencies of class or family position. But then neither should it depend on the natural lottery of talents, for talents are no more deserved than one's family background or starting place in the competition. Even one's character traits are not so clearly deserved, he thinks, for one's willingness to develop one's talents and use them for the benefit of others also depends in part on one's family background. Unless we are willing to do away with the family, then, it is very doubtful that we can secure equally good prospects for those who are equally well endowed, and hence, we need a principle 'which recognizes this fact and also mitigates the arbitrariness of the natural lottery itself' (Rawls, 1967, Section 6). The Difference

Principle appears to be a reasonable principle of this sort. Whether we are among the gifted or not, we stand to gain from the distribution of natural assets. 'Should we be less favored, we gain from the efforts of others; should we be more favored, we gain still more if we use our assets in socially fruitful ways' (*ibid.*, Section 10).

Quite clearly, many cooperative arrangements satisfy this weak condition. The competitive market, for one, and Hayek's 'each to benefit in proportion as he benefits those who have the resources to benefit those who benefit them' for another. The problem for Rawls is to advance considerations that single out the Difference Principle as the most reasonable principle meeting the weak condition cited.

Now, at the very least, fair terms of cooperation should be agreeable to all parties. And it is certainly plausible that the less fortunate would agree to the terms Rawls proposes. By offering larger distributive shares to the more fortunate, they increase their own absolute returns. Only an envious man desires to reduce another's larger share at cost to himself, and it seems entirely reasonable to exclude such unreasonableness (indeed, externalities generally) in arriving at a conception of justice. The more serious problem for Rawls is to show why the more fortunate, who would fare better in a market economy or a Hayek economy, would agree to his terms.

Most of us would concede that they should make gain from their greater gifts or better fortune only to the extent that they benefit others – though why, as Rawls insists, only the least favored? But this weaker condition is satisfied by a competitive economy. One needs a stronger condition, and Rawls invokes a Hobbesian consideration (attributed to Allan Gibbard) that 'the Difference Principle . . . seems to be a fair basis on which those better endowed, or more fortunate in their social circumstances, could expect the willing cooperation of others when some workable scheme is a necessary condition of everyone's welfare' (*op. cit.*, Section 10). This is a rather curious ploy. It seems intended to suggest that the less fortunate enjoy a superior threat advantage, but, in fact, it might be that the less fortunate would have more to lose by the breakdown of cooperation than the more fortunate. This is especially true if we consider the possibility of the better endowed cooperating among themselves and developing clever machines to do the more menial chores.[5] The weakness of this ploy may have prompted Rawls to introduce the maximin argument. If the cooperating parties are assumed ignorant of their talents and preferences, it might seem most reasonable of them to view the advantages flowing from cooperation and the distribution of assets from the standpoint of those who benefit minimally. Before this contractarian argument can carry any conviction, though, we must pin down the meaning of 'least fortunate'.

I think it would be well here to distinguish those who are *least favored* (i.e., least well endowed) from those who are *worst off* in a utility sense. Rawls has the former sense of 'less fortunate' uppermost in mind in the following very characteristic passage (Rawls, 1967, Section 4):

> It may be that the difference principle represents, in effect, an original agreement to share in the benefits of the distribution of natural talents and abilities, whatever this distribution turns out to be, in order to alleviate as far as possible the arbitrary handicaps resulting from our initial starting places in society. Those who have been favored by nature, whoever they are, may gain from their good fortune only on terms that improve the well being of those who have lost out. The naturally advantaged are not to gain simply because they are more gifted, but only to cover the costs of training and cultivating their endowments and for putting them to use in a way which improves the position of the less fortunate. We are led to the difference principle if we wish to arrange the basic social structure so that no one gains (or loses) from his luck in the natural lottery of talent and ability, or from his initial place in society, without giving (or receiving) compensating advantages in return.

As for the 'worst off', Rawls would have us compare the different representative positions by using an appropriate index of 'primary goods'. *Primary goods* include 'rights, liberties and opportunities, income and wealth, and the social bases of self-respect' (Rawls, 1975, Section III). Men are assumed to want these goods whatever else they want and whatever their final ends. The worst off are those whose portion of the available primary goods is a minimum (under an appropriate index).

As in the quoted passage, Rawls seems to *assume* that the least gifted (those who have the least marketable genotypes, given extant tastes and preferences) will be worst off under any fair terms of cooperation, but hopes to redress this morally arbitrary feature by ameliorating their condition to whatever extent possible. But notice, there is no necessity that the least gifted be worst off. Such a state of affairs might result from contingent facts about preferences, incentives, voluntary exchange, and the like, but it needn't. That is, we need an empirical premiss to the effect that the more gifted require larger distributive shares as a necessary incentive to exercise and develop their talents in beneficial ways. In this vein, Rawls recommends implementing the Difference Principle by superposing a minimally distorting redistributive tax on a competitive economy (Rawls, 1971, Section 43). The aim here is to compensate the less skilled by providing them the wherewithal to obtain more consumer goods and services than they would obtain in any alternative feasible economic state. At any rate, Rawls' own presumption that the least favored by nature will be least favored by the set of institutional arrangements is so strong that he scarcely notices the distinction between them. And yet, if we are troubled by allowing the distribution of primary goods to depend on arbitrary social contingencies, we should be troubled by a conception of justice which automatically assigns the worst off positions to those who fare worst in the natural lottery. Rawls is surely right in thinking that incentives are needed but it doesn't follow that a market economy is the only way to provide the needed incentives.

Even if we reject the presumption that the least gifted be worst off, other problems remain. Rawls does not specify an index (or weighting) of primary goods, and it is not easy to see how the parties could agree to the Difference Principle without knowing what index is to be employed, if shrouded by ignorance of what their skills and preferences are likely to be. About this Rawls himself remarks that 'the precise weights adopted in such an index cannot be laid down ahead of time, for these should be adjusted, to some degree at least, in view of social conditions' (Rawls, 1975, p. 96). This seems to concede that rational men will ordinarily differ in the relative importance they attach to different primary goods, although Rawls insists that *rational* men could never be willing to sacrifice liberty or equal opportunity for any other benefits. In *A Theory of Justice* (esp. p. 98f.), two such indices are tentatively proposed, but the important point is that Rawls regards an index of the proposed sort as given to the parties in the original position. But since (cf. above) any such index represents a weighting of the different primary goods, the choice of an index violates a plausible condition of individual sovereignty and renders the Difference Principle preference *dis*regarding. The other alternative would be to have the cooperating parties select the appropriate index (when the veil of ignorance is lifted), but it is not easy to see how that could be done without allowing a politically dominant group to tailor the index to their own advantage, and besides, the political institutions which govern their collective choices already reflect a weighting of the kind sought.

To see this more clearly, consider a two person economy whose members, Crusoe and Friday produce and consume the entire product. Supposing the fundamental liberties secured, the relevant primary goods appear to be leisure and consumption. But who is 'worst off' if, for example, Friday consumes more but also labors more? Only an assignment of relative weights to leisure and consumption can decide the issue. But the two parties may well differ in their assignment of relative weights, and since they are assumed ignorant of their tastes in the original position, the decision problem Rawls poses is ill-defined.

Let us return, then, to the other interpretation of the Difference Principle which permits larger distributive shares to be given to the more gifted (or socially advantaged) only when such bestowals improve the lot of those below. This interpretation disposes of the worry that the 'worst off' might prove to be a class of talented idlers, leading a Rawlsian society to conclude that some of the income of the working classes be redistributed to the idle poor! Rawls' own remarks about regulating the economy in the interests of efficiency suggest an interpretation of the Difference Principle that insures, not only mutual benefit, but that all parties *earn* their shares. But again, it is difficult to say precisely what Rawls has in mind.

Perhaps the following 'Rawlsian game of division' will help throw light on this matter. Suppose (as above) that Crusoe is the less productive (by endowment) of the two parties. By examining Friday's indifference map, or by trial and error, Crusoe determines the smallest share of their joint product at which Friday prefers to work an additional hour (or unit) when Crusoe's labor input leaves him s units of leisure. I refer to a labor input which leaves a man s units of leisure as *labor input s* in what follows. Let $p_t(s)$ be the smallest share that induces Friday to labor input t when Crusoe's labor input is s. When does it behoove Crusoe to offer Friday share $p_t(s)$? Obviously, when he increases his own absolute return by doing so, viz., $[1 - p_t(s)]\ P(s, t)$, where $P(s, t)$ is their joint production function and s is fixed. Let $p(s)$ be the distributive share for Friday (inducing him to give up a specified amount of leisure) which miximizes Crusoe's consumption when his labor input is s. Crusoe then selects the leisure-consumption pair which maximizes his utility among all the pairs $(s, [1-p(s)]\ P(s, t))$, where t is the labor input which share $p(s)$ induces from Friday.

This game seems a reasonable facsimile of the Difference Principle in operation. The distributive shares of the more favored (i.e., more productive) party are increased only when these shares increase the net consumption of the less favored party. The latter is allowed to adjust his own labor input and distributive share so as to insure himself the most preferred of all feasible leisure-consumption pairs. Intuition suggests that the higher Friday's productivity, the higher the distributive share it will behoove Crusoe to offer him. At the same time, the greater Friday's taste for leisure is, the higher the share Crusoe must offer to coax desired additional labor out of him. If Friday's substitution rate becomes too steep, however, he will price himself out of the market in the sense that it will no longer pay Crusoe to play the game at all (he would do better working alone and consuming the product). On the other hand, a potential for mutual benefit from cooperation (a *basis for cooperation*) exists if there is some division $(s, t; p)$ such that each party prefers his leisure-consumption pair to his optimal (feasible) such pair working alone (his *status quo*).

That rational interest requires one party to reward the greater productivity of the other with a higher distributive share is a consideration that applies symmetrically to both. But the Difference Principle invites us to view the advantages of cooperation from the perspective of the less gifted party (Crusoe). That he is allowed to select his own preferred labor input already biases the game in his favor. (On the other hand, to opt for maximum productivity subject to a distributional constraint is to bias the game against those who require much additional consumption to compensate them for the loss of a unit of leisure.) Even more serious bias enters in the fact that Crusoe grants Friday the *minimum* cut required to coax a desired labor input out of him. For each additional unit he works, Friday is gaining the minimum increment of product needed to make good his loss of leisure, given his rate of exchange between the two, while Crusoe will generally fare much better than that. It is easy to see, then, why Crusoe would agree to play the game, but less easy to see why Friday should. In a competitive market, a man can price his services, in light of the demand and alternative sources of supply, so as to maximize his net return. At the other extreme, society could price his services (assign wages), in light of the individual worker's preferences, so as to maximize its net return. (I.e., society would pay a man the lowest amount he would be willing to accept to do his preferred job rather than another.) The present interpretation of the Difference Principle results if we replace 'society', in the latter formulation, by 'the least gifted members of society'. It is unclear to me why one should be temped to endorse either of these corner positions. Let us seek a middle way.

4. Fair divisions

The Rawlsian game of division favors Crusoe (the less gifted party) by allowing him to hold Friday to minimum compensation for his loss of leisure, as measured by Friday's substitution rate, while garnering much better than minimum compensation for himself. Even if Friday is both more productive and harder working, he is apt to come out on the short end of it (utility-wise). What sort of division would two equally hard-nosed bargainers be likely to hit upon?

Suppose first that both partners are equally productive and that there is a basis for cooperation (a potential for mutual benefit). One way of insuring that both profit fairly from cooperation, neither taking advantage of the other, is to require that each be compensated with shares proportionate to the shares needed to compensate him for his loss of leisure. (I am assuming, of course, that neither party misrepresents his tastes.) That is a natural way of insuring that each make an equivalent sacrifice on the other's behalf and (since they are equally productive) that each confer equal benefit on the other.

Consider next the case where they have the same rates of exchange but differ in productivity. Then, clearly, shares should be proportioned to marginal product. We might measure Crusoe's marginal product by $P(s, t) - P(s_{\max}, t)$, the amount he adds to what Friday produces alone, and with a similar expression for Friday's marginal product. But, as I indicated in Section 1, the two may differ in the value or utility they attach to these production increments. If we are willing to countenance the relevance of their different leisure-consumption substitution rates, we should be willing to countenance that of the different utilities they attach to production increments. If they are equally productive, but one of them is more 'acquisitive' than the other (i.e., gains more utility from a given production increment), then they are *not* conferring equal benefits on each other, and, indeed, the onus seems to be on the more acquisitive man to compensate his partner with a higher distributive share (or, perhaps, with a leisure bonus). The corrected measure of Crusoe's marginal product is thus $U_F\,(P(s, t) - P(s_{\max}, t))$, where U_F is Friday's utility function.

Consider, finally, the general case where the parties differ in both their preferences and productivity. (I continue to assume that there is a basis for cooperation.) Again assume Friday is the more productive (in the extended utility sense), and let their respective labor inputs be s (for Crusoe) and t (for Friday). Then there is a labor input $t' < t$ for Friday at which $t' : s$ is equal to the ratio of Friday's marginal product to Crusoe's at labor inputs (s, t). (Recall that t, t' are actually amounts of leisure; that is why the inequality does not go the other way.) I call t' the *adjusted equal productivity input* (or *adjusted input*, for short). The idea here is simple enough: we equate the case where Friday is more productive (or, more accurately, where his labor contributes more to Crusoe's welfare than *vice versa*) with the case where he is equally productive but increases his contribution of labor just enough to make up the difference. This enables us to reduce the general case to the special case of equal productivities first considered. The crucial point, of course, is that Friday is now compensated on the basis of his adjusted input t' rather than for his actual input t. It is *as though* the two partners were equally productive but one sacrificed more of his leisure to the common good than the other. But notice, too, this theory pays no special respect to 'moral virtue' per se; from the present perspective, it is a matter of complete indifference whether the benefits one partner bestows on the other are owing to his extraordinary talents (his facility) or to his unusual dedication to the welfare of others. One may find this feature of the approach objectionable, but it does not, in any case, differ from the other conceptions we have been exploring in this regard.

I call distributions (or assignments of distributive shares) that conform with the above principle *fair*. The definition extends in an obvious way to *n*-person economies of our simple type. The concept of a fair distribution smoothly extends the qualitative principle of justice with which we began; when one of the relevant factors, marginal product or 'cost' to the contributor (as measured by his substitution rate) is held constant, the extended concept reduces to one of the two special cases the qualitative principle is able to handle. Fair distributions, as I have defined them, are also symmetrical in the required way. When two men agree to the proposed terms of co-operation, each profits in proportion as he benefits the other and in proportion to his own substitution rate for leisure and consumption. It is not enough that the terms of cooperation insure mutual benefit; the terms must be fair.[6]

5. Justice as Collective Choice

We must now consider ways of extending the concept of a fair division to more complicated situations where there are diverse economic functions to be performed by workers who vary both in the relative disutilities they attach to these different jobs and in the skill with which they perform them. Their preferences will be reflected in different leisure-consumption substitution rates for different jobs, these rates being steeper for less preferred jobs. That is, it would take more consumption (or a higher wage) to induce you to give up part of your leisure for work you abominate than for work you enjoy. More to the point, I will assume that the marginal decrease of utility is greater for the income from a more enjoyable or more interesting job, even though, at any fixed level of income or purchasing power, the utility of earning that income from a more enjoyable job will be greater.

The natural extension of my conception calls for distributive shares proportional to a worker's marginal product with due allowance for the costs he incurs, as measured by his labor input and substitution rate for the particular job he does. The only new element that bars this smooth extension is the fact that the different members of our imagined commune will differ in their acquisitiveness, in the utilities they attach to the commune's product. One could perhaps surmount this difficulty by some sort of averaging, but it will prove more satisfactory to follow another course. *I will imagine that the optimal fair division is determined by a rational collective choice procedure.* This more general conception of justice as the outcome of rational collective choice will be seen to underlie the conception developed in the preceding section.

Let us suppose that the members of the commune hire an omniscient economist who, by surveying the resources and skills available to them is able to list all feasible production-consumption schedules or economic states. The description of each such state includes: (1) the labor input of each worker, (2) the associated production schedule of goods and services, (3) the resources expended in production (including human resources), (4) the rate of saving, and (5) the distributive shares of each worker in that portion of their product destined for consumption. Investment is taken care of by combining (2) with (4). Diseconomies of production are handled under (3).

Each worker has an equal voice in the communal economic decision making. If we wish to exclude externalities, we may draw a veil of ignorance over the labor inputs and distributive shares of other workers in presenting the alternative economic states to a given voter. (The quality of given goods and services would *not* be veiled.) Ideally, the method of collective choice should reflect individual preferences and intensities of preference. It should be non-manipulable (so that no voter has an incentive to misrepresent his tastes), non-dictatorial, and non susceptible to collusion or voting in blocs. We could easily involve ourselves in a lengthy discussion of various methods of collective choice at this point, but I will circumvent such discussion by selecting an idealized method which equates votes with utilities.[7] More practical schemes will be considered in the sequel.

My more immediate objective is to illustrate the close connection between the divisions that result from collective choice and the 'fair divisions' of the preceding section. To this end, consider first a small business partnership whose three members, Al, Bob and Carl are, respectively, an excellent salesman, a competent accountant, and a skilled foreman. I will assume that all three have identical preferences as between the three functions to be performed: all three prefer sales and service to bookkeeping and the latter to labor relations and production. Label the three jobs 1, 2, 3, so that an assignment of labor inputs may be written $\binom{ABC}{ijk}$. Time-sharing of the same job would be considered in a comprehensive treatment, but we neglect such possibilities here in order to simplify the presentation.

Let x, y, $1-x-y$ be the distributive shares associated with jobs 3,2,1 (ordered from least preferred to most preferred) and let S be the total portion of product destined for consumption. The associated utility functions for the three jobs (common to A, B, C) are $\log(xS)$, $\alpha \log(yS) + \beta$ and $(\alpha-\gamma)\log(1-x-y)S + (\beta+\delta)$, where $0 < \alpha, \beta, \gamma, \delta < 1$ and $\gamma < \alpha$. Net utility is then given by $U = F(x,y) + (1 + 2\alpha - \gamma)\log S$, where

$$(9.1) \quad F(x, y) = \log x + \alpha \log y + \beta + (\alpha-\gamma)\log(1-x-y) + (\beta+\delta).$$

Since U is clearly increasing in S, labor is allocated so as to maximize production (hence, in our example, the labor inputs are $\binom{ABC}{123}$). The optimal distribution maximizes $F(x, y)$. Set

$$0 = \partial F/\partial x = \frac{1}{x} - \frac{\alpha-\gamma}{1-x-y}, \quad 0 = \partial F/\partial y = \frac{\alpha}{\gamma} - \frac{\alpha-\gamma}{1-x-y}.$$

Hence $1/x = \alpha/y$, or $y = \alpha x$. Plugging this into the first equation and solving for x gives $x = 1/(1 + 2\alpha - \gamma)$, whence $y = \alpha/(1 + 2\alpha - \gamma)$ and $1-x-y = (\alpha - \gamma)/(1 + 2\alpha - \gamma)$. Since $\alpha - \delta < \alpha < 1$, the result $x > y > 1-x-y$ follows. Note, finally, that $\partial F/\partial x > 0$ iff $(1 + \alpha)x + y < 1$. Hence U is increasing in x only so long as $(1 + \alpha)x < 1 - y$. Similar constraints govern the increase of U in y and $z = 1-x-y$. Subject to our assumptions, then, we have proved that the higher share of consumption is assigned to the less preferred job, and that labor is allocated efficiently (viz. so to maximize production). It is not hard to see, on the other hand, that labor inputs of the division determined by the sum-of-utilities criterion of choice need not maximize production when job preferences are allowed to vary from worker to worker. Notice, finally, that the assumed diminishing marginal utility of income (or consumption) will *quite generally* damp down inequalities of income, as illustrated by the constraints on the increase of U in x (or in y, or $1-x-y$) given above.

For a concrete illustration, take $\alpha = 0.8$, $\beta = 0.4$, $\gamma = 0.3$ and $\delta = 0.6$. The optimal distribution is then $(x,y,z) = (21.74, 34.78, 43.48)$ with associated utilities $(A,B,C) = (1.67, 1.63, 1.64)$ when the labor inputs are $\binom{ABC}{123}$ and this (optimally productive) set of labor inputs yields a product of 100 units to be consumed.

Our results obviously hold under more general conditions. When tastes are alike, there can be no basis other than their productivities for assigning one man rather than the other to a mutually preferred job. When tastes are alike, therefore, labor will be efficiently allocated so as to maximize production. With labor inputs set, only the problem of distributing the product remains.

At the optimum distribution, the marginal utility of income (or consumption) will be the same for all workers. For social welfare cannot be at a maximum if the last unit of consumption allocated to one worker nets him less marginal utility than it would have netted someone else. A worker whose utility per income curve is steeper than another worker's at their current consumption levels will generate more additional utility from a small increment of consumption. A sufficient condition that a worker receive a higher distributive share than another is that his marginal utility of income be decreasing less rapidly at every level of income, and we may regard that slower rate of decreasing marginal utility of income as definatory of a less enjoyable job.

There are several possible misconceptions worth pausing to clear away before proceeding to the problem of practical implementation. Although the idealized collective choice procedure equates votes with utilities, it would be misleading to equate the present conception of justice with a species of utilitarianism. I am not defending a utilitarian ethic, nor even suggesting that states of the world with higher net utility are 'better' (in some sense). Economic states not collectively chosen may well have higher net utility than some that are chosen. If collusive voting in blocs were allowed, for example, blocs with sufficient voting power could exploit or even enslave minorities, and it is at least conceivable that states in which a few are exploited to satisfy the many have higher net utility than states that would be collectively chosen when such collusive voting is disallowed.

Ironically, it is Bentham's own dictum, 'everybody to count for one, nobody to count for more than one', that goes to the heart of the difference between my conception and Benthamite utilitarianism. For the latter, as customarily interpreted at least, does allow utility trade-offs of the indicated kind. Bentham's dictum expresses a requirement of fairness or equality of treatment. When applied to the selection of an economic state it insures both that each man's preferences be accorded equal weight and that shares are distributed so as to secure mutuality of benefit. That is, utilities are *earned* in Rawlsian fashion. Those with high distributive shares have won them by benefiting others; those who choose to develop less socially valued skills pay what amounts to a 'social opportunity cost tax'. The more others would prefer economic states in which a given worker is doing another (dispreferred) job, the greater the tax he will have to pay. That is, those who have strong preferences for work they do not do especially well (as compared with other workers and as compared to other jobs) will have to sacrifice leisure or high wages or both to exercise their preference for rewarding work.

On the other hand, there is no presumption, as there seems to be in Rawls, that the less well endowed will receive smallest distributive shares. If relatively unskilled jobs are relatively less rewarding in themselves, but make indispensable contributions to the general welfare, those who undertake them will be awarded high wages or distributive shares. Since men who endure hardship, discomfort, boredom, or who risk their lives or health to render an essential service in peacetime are doing essentially what soldiers do in wartime, it seems entirely appropriate that society should accord such positions high honor rather than their more customary low status. And while it is true that absolute utility levels (as opposed to marginal utilities of income) may differ at the welfare optimum, the point to emphasize is that they will not vary in any patterned way. There is no predicting what levels of utility will be enjoyed by different classes of workers, much less how utility levels will vary from worker to worker within a given class of workers. Otherwise put, there is no 'worst off' representative position under the conception of justice as the outcome of collective choice.

Let me add here a more internal criticism of Rawls. He regards the social bases of self-respect as 'perhaps the most important primary goods' (Rawls, 1971, 440). Other primary goods he mentions, like liberty, opportunity, education, income and wealth are all regarded as goods largely because of the potential contribution they make to one's self-esteem. I do not wish to quarrel with any of this, nor with the importance Rawls attaches to the contribution to one's self-esteem made by the opportunity to use one's talents in ways that others perceive as beneficial. Yet, insofar as he discusses practical implementation at all, Rawls appears to regard a competitive market economy with income redistribution effected by expenditure taxes as the best way to implement the Difference Principle. I seriously doubt whether redistribution is the best way to build the self-respect of those at the bottom of the economic ladder, especially when one considers that, in a competitive economy, there is likely to be a high correlation between one's place in the natural lottery of talents and one's place on the economic ladder. Indeed, something like the proposed combination of competition and redistribution prevails at the present time without having notably enhanced the self-esteem of those who occupy the lower rungs. Far more self-respect is likely to result from a scheme which appropriately rewards those who, while less gifted, are disposed to tackle the jobs generally thought to be onerous or distasteful. Rawls would compensate them for their undeserved place in the natural lottery by offering the more gifted larger shares as an incentive to increase the living standards of those who are below. It is only if we presuppose a competitive economy, however, that we will incline to the view that the more gifted, who will generally be engaged in the more attractive pursuits, require higher distributive shares to induce them to utilize their talents for the benefit of others.

Let me carry the ad hominem drift of these remarks one step further. Our approach to justice through collective choice is reminiscent of the Rawlsian contract. Both conceptions view justice as the outcome of a collective choice by free and equal parties, but there are important differences as well. The output of our collective choice procedure is an actual division of burdens and benefits; that of the Rawlsian contract is a set of principles which shape the assignment of fundamental rights and responsibilities, as well as the political and economic institutions which influence the life prospects of those who occupy the different representative positions. The constraints on knowledge that characterize the original position are designed to prevent any of the parties from tailoring the principles of justice to his own advantage. Only mild limitations of this sort are needed under the voting procedure I have described, specifically, limitations designed to exclude externalities (envy, spite, favoritism, racial, religious, or sexual discrimination, and the like) and to prevent collusive voting in blocs. Now the idealized voting procedure I have described already seems to treat the parties as free and equal moral persons. Moreover, for just this reason, it is not implausible to suppose that rational beings would embrace this procedure as a fair basis for cooperation from behind the Rawlsian veil of ignorance, where each is ignorant of his skills and tastes. For the parties are assumed to be cognizant of the general laws of human behavior. I take it the diminishing marginal utility of income could reasonably be included among these laws. Consequently, those in the original position know that inequalities of income arising from our procedure are generally damped down by the operation of this law. They are also in position to recognize the reinforcement which the appreciation of others affords one in developing any skill or talent. Hence, they can assume that there will be an adequate demand for most of the services they will be disposed to render.

I could go on in this vein citing additional generalities, but I expect I needn't belabor the obvious: one would not be likely to lead an intolerable existence under the democratic method of determining the burdens and benefits of cooperation, whatever one's skills and tastes. On the contrary, the democratic method seems to avail each worker as fair a prospect of leading a productive and satisfying life as he could reasonably expect in an imperfect world. In particular, the democratic method compensates the less advantaged either by higher wages or more leisure (for doing socially dispreferred work), or by opening up more attractive jobs at lower salaries. This flexibility would not characterize a society governed by the Difference Principle, and the latter, as I emphasized above, cannot be implemented without adopting an index of primary goods which would risk biasing the system of rewards against given preferences, above all, the preference for more leisure or for more stimulating work. It is noteworthy that the word 'leisure' is nowhere to be found in the massive index of *A Theory of Justice*. Nor does that book manifest any concern with interesting work as such, a notable omission from Rawls' list of primary goods.[8]

6. Practical implementation

Because the conception of justice as the outcome of collective choice averts the difficulties associated with the notion of a least favored representative position, one would expect the lines of its practical implementation to be delineated more clearly. While not presuming to offer a comprehensive blueprint, and recognizing the risks involved, I will attempt to sketch here a number of practical schemes which partially realize the idealized voting procedure described above.

In some locales, real estate taxes are based on self-assessed property values, but with the understanding that anyone may purchase a property at its owner-assessed value.[9] Nicolaus Tidemann has suggested to me the possibility of applying the self-assessment device to wages, as follows. All the jobs or job categories are listed, and each worker states the minimum wage that he would accept to do each listed job. Workers are then hired at their self-assessed wages. (Workers not hired to do any job would presumably have to lower their sights; society could, if it chose, act as an employer of last resort.) For each job he is not hired to do, a worker must pay a social opportunity cost tax proportional to the wage he assigned himself for that job. In this way, a high minimum wage functions as insurance against having to do a dispreferred job (or else as adequate compensation, if one's services remain attractive at that wage). The tax is the premium one pays for this insurance, and society can adjust it to suit. Quite clearly, labor will continue to be allocated with reasonable efficiency in this system, for the best qualified of those asking the same minimum wage for a job will be first hired. Indeed, the self-assessment system has many market-like features, but it does, nevertheless, relate remuneration more directly to the distastefulness of certain kinds of work, both in bidding salaries up and in taxing those who are hired to do the more rewarding or enjoyable jobs. Part of the tax revenues could be used to finance job training for the less skilled. It also seems likely that more mobility and opportunity would obtain under self-assessment. In a competitive market, a job is advertised at a given wage. Workers compete and the best qualified are hired at the stated wage. But under self-assessment there is the attractive option of setting one's wage for a desired job sufficiently low to make one's services attractive at that wage. Nor is any productivity lost thereby, for society is still getting at least as good a return on its wage dollar. No doubt, the intriguing strategic considerations that arise under self-assessment would make the costs of calculation rather high for individual workers. Each would have to guess at the smallest wage that would price his services out of a given job market, and that would depend on the wages set by others. Still, the element of chance would appeal to a great many people. Long-term contracts would be necessary to insure that those who have undergone expensive professional training remain on the job long enough to justify society's expense on their behalf. Reasonable penalties could be charged for breaches of contract, and employers wishing to discharge a worker would pay a penalty determined by his length of service in order to insure against capricious firing or farming out of older workers. At the same time, workers who are frequently discharged would be penalized by reduced marketability of their services, just as in a competitive economy.

I consider now a second scheme, due in its essentials to Oscar Lange (cf. Lange and Taylor, 1938, pp. 100 ff.). All workers are hired into a floating labor pool at a standard wage, but any worker is free to purchase more personally satisfying work. The purchase price is proportioned to the social preference for that kind of work and inversely proportional to skill, which is here conceived as a blend of aptitude, relevant experience, and past performance. The social preference of given work is measured by what people are willing to pay for the privilege of doing that work. Where the supply of labor, at a given level of skill, falls short of the demand, the purchase price is lowered; where the supply exceeds demand, the purchase price is raised. As in a competitive market, the resulting wages are thus responsive to demand for a service and the scarcity of the skill needed to provide it (at desired levels of proficiency). But, again, the Langean system makes wages far more responsive to the distastefulness of the work than wages in a purely competitive economy. Since the purchase price of each job grows steeper with declining skill, labor would continue to be allocated with reasonable efficiency, especially as wage rates are raised or lowered in response to demand.

The Langean device is best viewed as an integral part of a more comprehensive social blueprint, market socialism. There is free choice of occupations and private ownership of consumer goods, but public ownership of the means of production. The problem for socialists is to insure both consumer sovreignty and rational allocation of resources (including labor). Building on the earlier work of Pareto's student, Enrico Barone, Lange and Taylor (1938) propound the following solution.

Imagine that prices of all products and factors are given. A worker has an income and an array of goods and services from which to choose. He is consuming at an optimum when the marginal utility of consumption is the same over all commodities and services purchased. Similarly, each firm has an array of production schedules from which to choose. It must determine the combination of factors to use and the amount of each commodity to produce. Technical efficiency is attained when the marginal productivities of all factors are equal, and production quotas are optimal when marginal costs of production are equal to prices (i.e., when the cost of producing the last item is equal to the price at which it is sold). Prices function here as fixed parameters which guide the economic behavior of producers and consumers alike. In practice, of course, shifts in supply or demand force an upwards or downwards adjustment of prices.

All of this pertains, in the first instance, to an ideally competitive economy characterized by open markets and by a multitude of sellers no one of whom controls enough of the market to influence the prices of products or factors. Prices discharge, then, a social accounting function, guiding production to the most efficinet use of available resources. Waste is indicated by oversupplies of some goods and undersupplies of others. Where shortfalls exist, current prices are ineffective guides to behavior and need to be adjusted. Now, as Fred Taylor first noticed, this adjustment of prices, which proceeds by trial and error in a competitive market, can be performed just as well by the managers of a socialist economy. It is only that some prices – those of the ultimate factors of production – are stripped of all but their social accounting function; they do not represent profits or losses to anyone. Indeed, no monetary transfers need occur when ultimate factors are allocated to the different industries. Each manager is directed to produce efficiently and to satisfy consumer demand; these two conditions determine an industry's demand schedule for the ultimate factors of production. Suppose, for example, that society wants to determine the most efficient use of land. The equilibrium prices of various food and wood products must first be estimated, and, using these, the equilibrium wages for the labor each use would require. Consumer demand and labor preferences are then reflected back to a demand for land, and it is decided how much land would be purchased at each price so as to maximize profits. Equilibrium prices for land are then found, and the available land is allocated as in a competitive market.

Industrial managers who follow the rule that the cost of the last item produced not exceed its price will show profits. But there is no need to insist that every concern show profits on an annual basis. The total revenue from sales of custom-made products, like works of art, need not cover the salaries paid out to artists and custom craftsmen in a year. But works of art are distinguished from most industrial goods by their tendency to appreciate rather than depreciate in value. This raises subtle accounting problems best left to economists. Without delving further into the matter, let us note the possibility of correcting the annual net revenue from art works by a factor that depends on their average rate of appreciation, and then basing annual salaries on the corrected quantity. Individual artists could also be paid bonuses or royalties for works sold or commissioned. In effect, the envisaged financing of the arts functions as a risk-sharing arrangement with built-in incentives to insure a degree of responsiveness to public taste; it has an obvious contractarian rationale.

A strong case has been made for market socialism on economic grounds. Roughly speaking, one can argue that market socialism combines the planning and coordination of oligopolistic capitalism with the economic efficiency of an ideally competitive market. Indeed, it can be considered more efficient that a competitive market insofar as it satisfies two additional desiderata: (i) the same demand price offered by different consumers represents equal urgency of demand, and (ii) the difference in the value of the marginal products of any two labor inputs or services is equal to the difference in marginal disutility incurred by those who provide those outputs or services (Lange and Taylor, p. 103). While competitive markets are technically efficient and utility efficient for consumers, they are not very utility efficient for workers. For the option of doing more preferred work at a lower wage is generally unavailable.

Moreover, because it is able to enter all costs of production on its accounting sheets and to consider all economic alternatives, a market socialist economy eliminates much of the economic waste associated with monopolistic forms of capitalism, among them, unemployment, diseconomies of production (like pollution), business cycles, and resistance to technological innovation.[10]

Given my narrower concern with economic justice, however, it is no part of my purpose to argue the case for market socialism in detail. It is enough to have illustrated the serious economic considerations that have been marshalled on its behalf.[11] If market socialism is unworkable, it must be for more subtle reasons. One must believe that a socialist society would lack discipline, that the managers would lack sufficient incentives to perform competently, or something of that sort. Market socialism is, above all, plausible socialism, and that makes it less easy to dismiss the conception of justice that underlies it as impractical fantasy. The socialist blueprint functions here as an existence proof.

7. Liberty

Some will grant that market socialism is efficient but will harbor reservations about its compatibility with individual freedom. Opponents of socialism labor under an ever-present temptation to load their definition of the term with overtones of authoritarianism, even of totalitarianism. In point of fact, there is a vigorous socialist counter to this charge which, not only denies the existence of a conceptual link between free markets and free societies, but questions whether democratic institutions can long remain viable when tied to a capitalist economy.[12]

John Rawls has also been concerned with the problem whether justice can be attained without sacrificing liberty. His insistence on the priority of liberty is based on a denial that free and rational parties, operating from behind the veil, would ever choose to sacrifice even the smallest amount of liberty for economic advantage. But Rawls' many discussions of this matter leave many questions to be answered.

His first principle of justice requires that 'each person engaged in an institution or affected by it has an equal right to the most extensive liberty compatible with a like liberty for all' (Rawls, 1967, Section 2). Consider the institution of contracts. If we separate freedom of association and the right to enter into contracts from the question of legal enforcement of contracts, it is hard to see what (if any) constraints Rawls' principle would impose. The liberty of two men to enter into a contract to murder a third is surely compatible with a like liberty for all. So, too, is the liberty of two firms to fix prices, or, for that matter, the liberty of two property owners to divert a stream that supplies water to other property owners. Liberties must be compatible with others' rights, and not merely with 'like liberties' of others. But even this requirement seems insufficient. Price fixing, for example, does not appear to violate anyone's *rights*, for there is no right to buy a desired commodity at a competitive market price (as libertarians would be the first to insist). It is not clear, though, whether free and equal parties operating behind a suitable veil of ignorance would consent to price fixing or other monopolistic practices. And the major thrust of Rawls' work, I should have thought, is to evaluate all institutions from a contractarian perspective. In line with this reading, Rawls does recommend regulating a competitive economy in the interests of efficiency (when discussing the implementation of the Difference Principle), and this conjures up anti-trust legislation. Consequently, if there is a liberty to fix prices, Rawls must be understood as recommending that *this* liberty be curtailed for the sake of economic efficiency. These remarks underscore the need to fill out and circumscribe Rawls' first principle. His overriding concern, no doubt, was to provide a contractarian basis for the fundamental political liberties (of thought, conscience, speech and press, etc.), and in this he achieved a measure of success. But his discussions contribute little to the clarification of the issues that arise when one man's actions result in a narrowing of choice for another. I will take up this matter in developing the main line of discussion. Let us return then to the question whether a market socialist economy could long endure without sacrificing liberty.

We are aided here by the recent work of Robert Nozick (Nozick, 1974) which poses this issue in a specially incisive way, culminating in the charge that 'the socialist society would have to forbid capitalist acts between consenting adults'. Let me quote more of his argument (Nozick, 1974, pp. 163, 168, 172):

> . . . no end-state principle or distributional patterned principle of justice can be continuously realized without continuous interference with people's lives. Any favored pattern would be transformed into one unfavored by the principle, by people . . . exchanging goods and services with other people, things the transferrers are entitled to under the favored distributional pattern . . . .
>
> Patterned principles of distributive justice necessitate *re*distributive activities. The likelihood is small that any actual freely-arrived at set of holdings fits a given pattern; and the likelihood is nil that it will continue to fit the pattern as people exchange and give. From the point of view of an entitlement theory, redistribution is a serious matter indeed, involving, as it does, the violation of peoples' rights . . . .
>
> Taxation of earnings from labor is on a par with forced labor . . . .
>
> End-state and most patterned principles of distributive justice initiate (partial) ownership by others of people and their actions and labor. These principles involve a shift from the classical liberals' notion of self-ownership to a notion of (partial) property rights in *other* people.
>
> Considerations such as these confront end-state and other patterned conceptions of justice with the question of whether the actions necessary to achieve the selected pattern don't themselves violate moral side constraints.

This scattered quotation from Nozick puts the difficulty in sharp focus: to realize a favored pattern of distribution or end-state principle of justice (and Nozick includes the principle that proportions distributive shares to marginal product under that rubric) necessitates continuous interference with voluntary exchanges and continuous redistribution of goods to nullify such exchanges. In what sense, though, is the conception of justice as the outcome of collective choice (or the mechanisms to realize it described in the preceding section) 'patterned'? How are the distributions which result from individual choices under market socialism any more or less 'patterned' than those which result from individual choices and exchanges in a competitive market? If anything, they differ only in the greater complexity of the patterns generated. In a competitive market, the 'value' of a service is a joint function of the demand for that service and the scarcity of the talent needed to provide it. These factors remain determinative of value under market socialism and the self-assessment scheme as well, but another factor, the distastefulness of the work, is also constitutive of value – at least to a greater degree. There are a great many people who can collect the garbage – the talent is not scarce – but relatively few may be disposed to take on that task, and hence the value of the service will be increased. The wages of distasteful work will be bidded up in a competitive economy too, but only to a very slight extent (or not at all) if there is a large reserve force of unskilled laborers. Indeed, the tendency will be to replace unskilled labor by machine labor, and thereby further drive down the market value of unskilled labor.

No more is there any reason to qualify the actions and decisions taken in a competitive market economy as more 'voluntary' or 'free' than those taken under market socialism or the self-assessment system of wages. In both cases the individual confronts a constellation of alternatives which blend different amounts of job satisfaction, purchasing power, and leisure. In both cases wages function as parameters of action, as conveyers of pertinent information. The information conveyed varies with the economic system. In a perfectly competitive economy, wages are indices of the demand for a service and the scarcity of the skill needed to provide it. In a market socialist economy, wages will convey additional information about social preferences for different kinds of work. I fail to see, however, what could lead one to characterize the decisions taken in the one case as 'free' and those taken in the other as 'forced'.

If anything, there is more 'positive freedom' (a wider range of choices) under a socialist or self-assessment economy than under a purely competitive economy. The latter has been known to place people in a position where their 'choice' is to labor more than eighty hours a week for a bare subsistence wage or starve. I say this 'choice' is no choice and their putative 'freedom' is no freedom. The theory of the voluntary which underlies this claim derives from Aristotle's discussion of 'mixed cases' in Book I, Chapter 1 of his *Nicomachean Ethics*, cases like throwing a cargo overboard in a storm to save a ship, or divulging military secrets under threat of torture. Such actions, he says, are 'more like the voluntary, for they are worthy of choice at the time they are done'. But, farther on, he adds that 'in the abstract, they are perhaps involuntary, for on one would choose any such act for its own sake'. He then throws out an important clue to their correct classification when he says: 'On some actions praise is not bestowed, but pardon is, when one does what he ought not under pressure which overstrains human nature and which no one could withstand'. Hence, to admit that anyone with the same natural characteristics would have acted as the agent did is to absolve the agent of blame. But where the agent is not to blame, I would add, he is not (morally) responsible; his action should be accounted involuntary. And, moreover, it is then rather odd to say that he ought not to have done it when, under the circumstances, he could not help but do it. Aristotle's 'mixed cases' are thus perhaps best thought of as 'mixed' in the sense of being neither wholly voluntary nor wholly involuntary. Rather, they are partially voluntary actions (voluntariness admits of degrees), and degrees of voluntariness are measured, not metaphysically, but statistically, as the proportion of those (with the agent's natural traits or no-fault traits) who, under the circumstances, would have acted as the agent acted. To say that he *could* have done otherwise, therefore, is to say that most others *would* have done otherwise. Since, pretty clearly, most would work twelve hours a day in a mine rather than starve, my theory accounts their actions as (largely) involuntary and their freedom as rather minimal (supposing they have no more attractive alternatives). The line between narrowing of choice and outright coercion becomes razor thin, perhaps academic, in such contexts. Asked why he is working that hard for so small a wage, a man could very naturally reply 'I have no choice'.

Nozick takes a very different tack here, arguing that when the actions of others limit my options, my subsequent action is voluntary or not according as they acted within their rights or not (Nozick, 1974, pp. 262 ff.). Whenever anyone else comes to possess something – property, a wife, a concert ticket – my options are narrowed, but my subsequent actions are voluntary if he was within his rights to acquire what he did. Now according to Nozick's 'entitlement theory of justice', voluntary transfers are ipso facto just, and one may justly acquire property (though he never gives us a full blown theory of just acquisition). Any holdings which result from just acquisitons via voluntary transfers are accounted just. In particular, if these face me with the option of laboring eighty hours a week at a subsistence wage or starving (while others luxuriate), my choice of the former now counts as voluntary.

Now I should think that giving up something in one's possession at gunpoint would be a paradigm of involuntary action or coercion. But the person holding the gun might be either a policeman (who is acting within his rights – nay, as duty prescribes) or a thief. Hence, Nozick's theory makes the voluntariness or involuntariness of my action depend on who is holding the gun (supposing that I am on my way out of a bank with a heist). That seems very strange to say the least. Only those interested in forging a conceptual tie between freedom and a market economy could be tempted by an account like Nozick's. For my part, I feel no hesitation in saying that where prior transfers and acquisitions place anyone in a position where he must submit to exploitation or starvation, those acquisitions or transfers were unjust. Certainly economic systems that lead to such results with high probability are highly impugnable from a contractarian point of view.

Every exchange in a market economy is 'voluntary' in the Pickwickian sense that the parties preferred making it to not making it. But we cannot abstract from the alternatives, the matrix of choice. One of the parties may have enjoyed an unfair threat advantage, an advantage to which he was not entitled on contractarian grounds. That is precisely what happens when a dominant firm buys out its few remaining small competitors and becomes a monopoly. The dominant firm enjoys an unfair threat advantage; it can undersell its competitors or win any price war. Faced with that threat, the 'choice' of selling or being driven out of business is, again, no choice (whence the deliciously ambiguous 'He made me an offer I couldn't refuse!).

But wouldn't a socialist economy necessarily interfere with voluntary exchange? Wouldn't it be compelled, as Nozick alleges, to 'forbid capitalist acts between consenting adults' or forcibly restrain a man from selling his services to the highest bidder? My answer is that it would not! The highest bidder would be society itself, or its appointed bargaining agent. And society would pay each man a fair wage, as determined by the idealized collective choice procedure described earlier. Nobody would be tempted to say that a competitive economy limits a man's freedom merely because others refuse to buy what he is selling at the price he is asking. But then why should anyone be tempted to think this about a member of a socialist economy who wants higher wages than the collective decision making process accords him? Do we want to say that a man's freedom is limited merely because he is not able to collect the wages he thinks he should receive, or the wages he would receive under an alternative (and presumably unjust) arrangement? The claim that his liberty is infringed could not even be sustained on Nozick's theory of the voluntary, for it would have to be shown that society is not within its rights to fix wages by the Langean (or the self-assessment) device. To simply assert this would be, quite patently, to beg the question.

James Buchanan[13] varies Nozick's celebrated Wilt Chamberlain example in ways that sharpen this point. He imagines that Wilt lives on an island off the coast where he is unable to market his athletic skills, and where his alternative prospects are worth a mere $1000. If his skills would fetch $250 000 on the open market on the mainland, then he should be willing to pay up to $249 000 for an immigration visa, leaving the mainlanders to divide that entry fee among themselves as they see fit. Buchanan then continues:

> Let us change our earlier variation on Nozick's example to allow Chamberlain to be a member of the community from the outset. Now suppose that, through a wholly voluntary set of agreements, all prospective purchasers of Chamberlain's services appoint a single agent to deal with Chamberlain. As a result he agrees to provide services for, say, $10 000, with the rental value of $240 000 available for general distribution to all consumers. What would be Nozick's reaction to this end result? There has been no coercion; hence, he would have to agree that the holdings . . . are just. But this set of entitlements would be quite different from those which suggest Chamberlain gets $250 000 . . . . Surely if end-states as widely variant as the two we have postulated can be reached by fully acceptable processes, the content of the entitlement theory of justice seems minimal.
>
> How would Nozick respond to this charge? He could hardly call upon the forces of the minimal state to coerce prospective purchasers in such a way that the appointment of a single bargaining agent would not be allowed.

On the other hand, he might, for in the passage I quoted above Nozick charges that end-state principles 'involve a shift from the classical liberals' notion of self-ownership to a notion of (partial) property rights in other people'. It becomes apparent in such passages that Nozick's real case against socialism is based, not on its alleged unwarranted interferences with freedom, but on a very strong theory of property rights, one which holds that people have an entitlement to collect the entire rental value of their talents. (I cannot think what else Nozick means by 'self-ownership' in this passage.) No doubt, a man has a perfect right to deploy his talents as he sees fit, just as society has a perfect right not to reward him for rendering services its members do not value. But to have claims on others, one must enter into co-operative relations with them, and the terms must be acceptable to all concerned. Nozick seems to forget that the market value of one's talents is heavily dependent on the contributions of others. Athletic skills would not fetch very much without people to build sports arenas, transportation systems, and without the broadcasting and news media. By ignoring the gains from social interdependence, the case for 'self-ownership' in Nozick's strong sense is made to seem much stronger than in fact it is.

NOTES

[1] Would this violate the principle of 'equal pay for equal work'? It all depends on what counts as 'equal' work.

[2] As Oscar Lange observed (Lange and Taylor, 1938, p. 102), a socialist society might pay those with highly marketable skills incomes far below the marginal products of their services without reducing the supply of those services.

[3] This is a special case of a proposal I first heard from Patrick Suppes.

[4] Cf. Hayek (1960), Chapter IV, and Friedman (1962), p. 195.

[5] The 'Hobbesian' interpretation of the Difference Principle is further developed in Buchanan (1976a). Hobbesian anarchy is there taken as the benchmark to which the various feasible social states are to be compared, Buchanan argues that the Difference Principle would be chosen because the less favored could always undermine any other arrangement less advantageous to themselves and reintroduce Hobbesian anarchy by the simple expedient of withdrawing their cooperation. But, it seems to me, the same could be said of the more favored. For the possibility of the more favored cooperating among themselves, see Nozick (1974), pp. 193 ff.

[6] One might impose here the additional demand that each party *benefit equally* from cooperation. This could be construed as the requirement that each party increase his utility (over that which he enjoys in Hobbesian anarchy) by a like amount (cf. Gauthier, 1978), or better, as the requirement that each party have a margin proportional to his margin working alone, where a man's *margin* is the amount of consumption or purchasing power he actually enjoys less the amount needed to make good his loss of leisure, given his substitution rate. 'Bare subsistence' might be equated with life at the zero margin where one is producing only enough to cover one's 'costs'.

[7] For a collective choice rule that satisfies many, if not all, of my requirements, cf. Tideman and Tullock (1976). In their procedure, which develops an earlier idea of Edward Clarke's, each voter states how much he would pay to receive a given option rather than his least preferred option. The dollar votes for each option or candidate are summed and the one with the largest sum is elected. A voter who casts a decisive vote (one which alters the outcome) pays a *Clarke tax* proportional to the amount which his vote has cost the other voters, viz. the dollar vote for the option they would have chosen less that of the elected option, when the given voter's dollar votes are subtracted. The Clarke tax makes it unprofitable for a voter to misrepresent his tastes.

[8] If, however, one interprets the Difference Principle in a utility sense, equating the 'worst off' representative position with that to which minimal utility accrues, then, as J. D. Sneed has shown (Sneed, 1974), under rather mild conditions, the Difference Principle can be shown to recommend an equal utility point (a distribution in which all representative positions receive equal utility). The point, of course, is that the utility comes in different forms. For some, it assumes the form of higher purchasing power

(given high productivity), while for others, it assumes the form of more leisure or more interesting work. So interpreted, the Difference Principle comes much closer to the conception of justice as the outcome of rational collective choice.

[9] The dues levied on ships passing through the sound at Elsinore represent the earliest application of the self-assessment device known to me. Each ship's captain was permitted to assess the value of his own cargo, but the Danish king reserved the right to purchase any cargo at its assessed value.

[10] In capitalist economies, there is a need to preserve the value of invested capital, and a resultant tendency to resist innovation (e.g., by buying up patents of new inventions which threaten to render older methods of production obsolete). There will be a temptation to influence the government to adopt protectionist or restrictionist policies. No little peculation will be involved, and a government that rigorously supports unrestrained competition will run the risk that large firms will fail and send the economy into a disasterous tailspin. Rather than face that risk, government will tend to adapt its policies to the needs of the industrial system. It will serve those needs by regulating unemployment and aggregate demand (or consumer purchasing power), using the approved Keynesian methods: graduated taxation and investment. For these measures to have their desired effect requires the existence of a large, permanent public sector that will support technological development. National defense and the aerospace industry have answered these needs in the past. The result of these policies has been the growth of large firms subsidized by the government with consequent erosion of competition. There is further distortion of the market in the disappearance of alternatives. The development of jet aircraft, for instance, was financed largely by government defense contracts, but railroads were not similarly underwritten (that would have been socialism!), with the result that railroad passenger service all but disappeared for a time, despite the great need for good commuter services fostered by the growth of suburbs. For more on these matters, see Lange and Taylor (1938), pp. 111 ff., Galbraith (1967), esp. Chapters XIX, XXXI, and Schumpeter (1962), Part IV.

[11] Even Joseph Schumpeter, an avowed and persuasive champion of oligopolistic capitalism, is constrained to admit that the 'socialist blueprint is drawn at a higher level of rationality'. See his passionately dispassionate comparison of blueprints in Schumpeter (1962), Chapter XVII.

[12] See Note 10.

[13] Buchanan (1976b), pp. 5 ff.

BIBLIOGRAPHY

Barry, B., *The Liberal Theory of Justice,* Oxford University Press, Oxford, 1973.

Buchanan, J., 'A Hobbesian Interpretation of the Rawlsian Difference Principle', *Kyklos* 29 (1976a), 5–25.

Buchanan, J., 'The Libertarian Legitimacy of the State: a Critical Interpretation of Robert Nozick's *Anarchy, State and Utopia*', Working Paper, Center for Study of Public Choice, Virginia Polytechnic Institute and State University, 1976b.

Friedman, M., *Capitalism and Freedom,* University of Chicago Press, Chicago, 1962.

Galbraith, J. K., *The New Industrial State,* Houghton-Mifflin, Boston, 1967.

Gauthier, D., 'Rationality and Fairness in Cooperative Games', this volume, 1978.

Hayek, F. A., *Planning and Economic Order,* University of Chicago Press, Chicago, 1978.
Hayek, F. A., *The Constitution of Liberty,* University of Chicago Press, Chicago, 1978.
Lange, O. and Taylor, F., *On the Economic Theory of Socialism,* University of Minnesota Press, Minneapolis, 1938. Reprinted by McGraw-Hill, New York, 1964.
Nozick, R., *Anarchy, State and Utopia,* Basic Books, New York, 1974.
Phelps, E. S. (ed.), *Economic Justice,* Penguin Books, Hammondsworth, U.K., 1973.
Rawls, J., 'Justice as Fairness', *Phil. Rev.* 67 164–194.
Rawls, J., 'Distributive Justice', in P. Laslett and W.G. Runciman (eds.), *Philosophy, Politics and Society,* third series, Blackwell, Oxford, 58–82, 1967. Reprinted with addenda in Phelps (1973), and my references are to this version.
Rawls, J., *A Theory of Justice,* Harvard University Press, Cambridge, Mass., 1971.
Rawls, J., 'A Kantian Conception of Equality', Cambridge Review, 1975.
Schumpeter, J., *Capitalism, Socialism and Democracy,* 3rd ed., Harper and Row, New York and Evanston, 1962.
Sneed, J., 'Notes on a Theory of Justice', manuscript, 1973.
Tidemann, T. N., and Tullock, G., 'A New and Superior Process for Making Social Choices', *J. Pol. Econ.* **84** (1976), 1145–1160.

FREDERIC SCHICK

TOWARD A THEORY OF SOCIALITY

That people do what they do because of what they see in others hardly needs to be said. Nothing is more obvious. Still, the dependence of people on others calls for some analysis. I will show that two levels of dependence can be distinguished. I will go on to suggest that the second level is the one to study, and how this might be done.

The first level of dependence I call *responsiveness.* This is the dependence of people's interests on the interests of others. A person's interests are *responsive* if he has some of the interests he has only because he thinks that someone else or some group of people have interests of this or that sort. Putting it conditionally, a person's interests are responsive to those of some other person or group where, *if* he attributed these or those interests to these other people, his own interests would not be what they are. A person's interests are responsive to others' where they are a function of what he thinks are theirs. In such cases, we can speak also of this person himself as responsive to these other people.

This applies however we set out a person's interests. There are responsive desires. There are likewise responsive preferences and responsive utilities. However interests are represented, various patterns are possible. Adam's preference for p over q may be responding to Eve's preference for p over q, or to her preference for q over p, or perhaps to some preference of Eve's on some logically separate matter, or to some conjunction or disjunction or whatever of her preferences, or of her's and others'.

Certain patterns of responding merit special attention. This was stressed by Butler in the eighteenth century. Butler held that the fact that people pursued their interests was a mere tautology. Whatever interests people pursued would be theirs. On Butler's view, the point to note is that some people take the interests of others into account in establishing theirs, and that different people who do this do it in different ways. An altruist does it one way, a spiteful person another way, and there are other possibilities besides. If we simplify somewhat, we can map out a person's responsiveness in terms of an additive responsiveness function and build in weight-parameters marking this person's specific reactions to each of the others. On this representation, the difference between altruism and spite and the various other forms of responsiveness can be read in terms of the values taken by these

Hooker, Leach, and McClennen (eds.), Foundations and Applications of Decision Theory, Vol. II, 121–130. *All Rights Reserved*

parameters. An altruist weights the interests of others positively, a spiteful person weights them negatively, etc. (Better perhaps: a person is altruistic *toward another* if he weights *that other person's* interests positively, etc.)

The assumption of responsiveness helps to sustain the idea that people pursue their interests. In his remarks at the conference, Rapoport argued that the prisoners'-dilemma experiments proved that people are not always rational. Among his own subjects, there were those who were. He called these the *Q*-people. There were those who were Pareto-optimizers, people who always chose that one of their options that allowed for a Pareto-optimal outcome – these he called the *P*-people. There were also those who were competitive, people who always moved for the maximum possible difference between the payoffs to them and the others – these were *Z*-people. It might be held that the *P*-people and the *Z*-people were responsive to their respective opponents. On this view, the *P*-people were altruists – each assigned the interests of his opponent some positive weight. The *Z*-people were spiteful – for each of them this weight was a negative number. Butler's reply to Rapoport would have been that the *P*-people and the *Z*-people were fully as rational as the *Q*-people. They all pursued their interests. The only difference was that the *Q*'s were indifferent to the others and the *P*'s and the *Z*'s were not.[1]

The mention of Butler suggests that these matters have some bearing on ethics. Only two people figure in a prisoners' dilemma. Where there are others involved, each has a number of responsiveness parameters. How should the various parameters relate to one another? The classical utilitarians held that they should all be equal (and positive). Each person should count for one (or any other positive number), and no one for more than one. Less even-handed ethics have been set out too.

The concept of responsiveness also bears on political ethics. There is a long tradition of stressing the distinction between real and spurious, or true and false interests. On the classical view, a person's true interests are those that are not based on any false beliefs. They are those that are not conditioned on any error on his part. His false interests are those that are conditioned on error. If people's interests are simply collated in some way to establish the group interests, the result is almost certain to reflect many false interests. This cannot, in general, be considered a weakness of the collation procedure. The failings are those of the people involved.

But suppose that some of these people's interests are based on how they see others, and that it is these responsive interests that are false. I say that the situation with regard to these interests is in these cases not *in equilibrium,* and this in the usual sense – if everyone discovered where the others stood, at least one person would move. Given the possibility of a disequilibrium, the usual line that a democracy must respect people's actual interests loses some of its force. If equilibria are desirable, then interests that are false in the other-regarding respect may have to give way. There are all sorts of problems here, but I have discussed these in other places and so will hold off here.[2]

So much for responsiveness, for cases in which Adam's interests are dependent on Eve's. I want to move to a second level now, to cases in which it is Adam's *choices* that are dependent. Where Adam *chooses p* only because he thinks that Eve wants *p* or that she wants *q* or whatever, I shall speak of *sociality.* A person chooses *socially* where he chooses as he does only because he thinks that someone else or some group have these or those interests. Or again, as above, to put the idea conditionally, a person is socially dependent on others where, *if* he attributed these or those interests to these others, he would not now be choosing as he is. A person's social choices are a function of what he thinks the interests of others.

This may seem to be making much of very little. Where a person's interests with regard to some propositions satisfy certain conditions, his *choice set* can be defined in terms of these interests – it is the set of those options which are either preferred or indifferent to each of the others. So it may seem that no sharp distinction can be drawn here. Whatever we want to say about choices must be expressible in terms of interests.

This would be confusion. Choices can be made where the existence conditions for choice sets do not hold. More to the point, choice sets are not choices, they are the sets of those of our options that we should (or would) in certain conditions choose. The concepts of choices and of interests are distinct. What we say about choices is thus logically clear of what we say about interests.

Some critics will persist. They may argue that where we want some one of our options – where we have *settled* on it – because of the interests of others, we must at some time have *come to* want it because of them. In the context of an issue, to *come to* want something is to choose that something. Thus a responsiveness of desires with regard to the options implies sociality. Again, in fact if not in logic, our distinction breaks down.

This misfires too. The argument confuses coming to want something for a certain reason with coming, for that reason, to want it. It identifies coming to want-something-because-of-others with coming-to-want-it because of them. Responsiveness of desires with regard to the options implies the former. Sociality however only enters with the latter.

Granted, there often is a connection. Suppose that a person chooses in the light of some assessment of the consequences, and that his interests are responsive in this regard. Here responsiveness does imply sociality. But a person may look to others in a way that takes no account of what might follow. His dependent choices may be imprudent. Here is sociality without prior responsiveness. Our distinction presupposes that people do not always weigh the results, that they do not always pursue their interests. What Butler thought a tautology must be rejected. I am willing to let the tautology go.

The thesis that people sometimes choose socially does not conflict with the view that they always look to the consequences. It does, however, provide for going beyond it. It permits us to credit choices we could not have credited otherwise, indeed to make sense of such choices. Sociality involves a distinctive sort of reasons, and so provides for a distinctive sort of explanations. A person has a social reason where he is moved by the thought that someone or some group has these or those interests. We in turn have a social explanation where we note such a reason.

This last will have to be qualified. Not every consideration of the consequences makes for an adequate consequential reason. Where we think that a person looked to the consequences, we do not let the story stand unless we see how his choice was rational in his lights. So also, not every response to others makes for an adequate social reason. Suppose that a person, content with his job, decided to move to another on learning that some recruiting employer wanted to hire him. We could make no sense of this person. Why is he so attentive to the wishes of an unknown employer? Our theory of rationality sets out the formal features of what people look for when they look to the consequences. We have also a theory of bonding, which identifies the relations between people that typically make for social choices of this or that sort. The person described is choosing socially, but not in the context of any appropriate social bond between himself and the other. If he were yielding to his wife, there would be no problem. We could accept his awareness of her wishes as determining his choice, for husbands often defer to their wives in such matters. But people rarely defer to total strangers.

Let us get this into a sharper focus. Where a person chooses socially, his choice is a function of what he thinks the interests of others. But whether or not he chooses socially, his choice dispositions collectively can be expressed as a *social-choice function*. Specifically, they can be expressed as a function whose domain is the set of all couples composed of some issue he might face and some possible combination of interests of each other person with regard to every proposition, and whose range is the set of all the subsets of the propositions constituting the issue involved, the empty subset excepted. A person's social-choice function identifies, for every issue he might face and for every possible combination of the others' interests, that subset of his options to which this person's awareness of these interests would restrict his choice – or to which it would turn him back, if he already has chosen. Call this subset of his options his *residual set* in the given context. Where this set is a singleton, an awareness of the interests of others is conclusive for this person. Where it contains several propositions, other considerations might still be raised. Where it is the total set of this person's options, an awareness of the interests of others does not narrow the issue at all. These last are cases in which this person will not choose socially. In the others, he will socially choose the disjunction of the options in his residual set.

Note that sociality need not preclude reflection on other grounds. Where a residual set is not a singleton, a person may go on to weigh the options in the disjunction he is left with, and here his reasoning will no longer be social. Note also that sociality might involve only a sensitivity to the interests of others with regard to the propositions that constitute the agent's issue and their disjunctions. Indeed, in most cases, a social-choice function has a smaller effective domain than that in our definition.

Now for the theory of bonding mentioned above, the theory of the social-choice dispositions prevalent in our society, or in whatever subgroup interests us. This theory can be laid out as a set of generalized conditionals. The antecedent of each conditional reports some pattern of possible relations holding between the chooser and some (or all) the others, perhaps along with some specific facts about these others, such as that this one is old and that one very young and this one related in this or that way to some of the rest. The consequent ascribes some social-choice function to the chooser, and may be quantified in some way. Let me call a generalized conditional of this sort a *posit*.

A posit for certain (archaic) groups might say that, for all A, B and C, if A is a child of B and C, where the issue is that of marriage, A chooses as he thinks B and C wish. Posits for all groups will say that children defer to their parents on many issues, and parents on some issues to their children, that husbands defer to their wives on many issues, and wives to their husbands. (Which issues these are will differ from family to family – hence the need for quantifiers.) There will often also be posits marking the deferences linking people related in less intimate ways: as teacher and student, as doctor and patient, as object of past favors and source of them, as friend and friend. Note that posits are always general. They are set out without individual names, not even the issues being pegged to specific agents. All that comes in are variables, and descriptions of how the agent, whoever this is, is related to the others – along perhaps with some further characterization of these others.

The philosophical significance of all this derives from the fact that the concept of sociality is central to that of morality. I spoke above of the morality of responsiveness. My view is that responsiveness is in fact morally marginal. Morality, as I see it, is basically a matter of commitment and (on another level) of generosity, and these two concepts only make sense in the terms of sociality. This calls for some argument, which I cannot offer here.[3] I will simply assert here that every ethics can be expressed as a set of posits. The ethics of this or of that society or of this or of that period, or the ethics of this person or of that one, are expressed in those posits which the people in this society or in this period, or the specific persons noted, *want* to be true. Moral behavior is *social* behavior of a certain sort – this in our sense of sociality. It is that sort of social behavior which people wish were generally current. We might also speak here of posits recast from the indicative into the imperative. These I shall call *precepts*.

This theory of ethics is totally nonpartisan. It allows for every conceivable sort of position. At one extreme is *slavishness*. This might direct us always to choose as some special person or group (thought to take a stand on each issue) wants us to choose. Here we would not be left with any residual choice. I know of no ethics that goes all the way in this direction. The other extreme is *libertinism*, all of whose precepts identify the entire set of a person's options as his residual set. Nietzscheanism is a version of libertinism. All the typical ethical positions fall in between. (I have spoken of utilitarianism in connection with responsiveness. It can be read in our present terms too as an ethics proper of the medial sort.)

There is more to be said on this. Granted than an ethics can be expressed in terms of posits. Not every set of posits that any person or group might wish were true would identify an ethics. Suppose that, after the apple incident, Adam wished that husbands always chose in some way contrary to their wives' desires. 'Always (or always in these or those issues) choose the opposite of what your wife wants' does not look like a moral precept. Only certain sorts of precepts qualify as moral. The question is: what sorts of precepts are these?

We might look for a general answer, hoping to discover principles that apply whatever a precept's antecedent. Here our question reduces to: what conditions must every consequent choice-function satisfy? The Adam example brings out one condition. The choice function must be *non-oppositional* –

may not, whatever the issue, have the agent choosing something only if he thinks some other person wants the contrary. We might supplement this along quasi-Kantian lines. Recall that a situation is in disequilibrium if, where everyone thought that it held, at least one person would change his stand. A disequilibrium is *irremediable* where no sequence of such changes could yield a situation not in disequilibrium. The quasi-Kantian condition is that a choice function must be *in every way publicly generalizable* – that it may not, whatever the issue, be such that, were all to choose in conformity with it, an irremediable disequilibrium might be established. Kant's principle is: Act only on such maxims as you might want to become universal laws. Our principle is: Choose only in accordance with such precepts as all might follow without any risk of irremediable disequilibrium.

A choice function must be *in every way* publicly generalizable. The underscored clause is essential, for every choice function is publicly generalizable in some way in some groups. Whatever the function, if the group is sufficiently large, where the deference relations are laid out in an inclusive cycle – *A* deferring to *B, B* to *C, C* to *D* and *D* to . . . *A* – joint choice in conformity with that function is publicly generalizable. But for the in-every-way clause, every function would pass the test in groups of a certain size. An alternative version of the principle is that a choice function must be *in every group* publicly generalizable – that it may not, whatever the issue, be such that, were all the people in any group (however small) to choose in conformity with it, an irremediable disequilibrium would be established. (Note that 'might' becomes 'would' here.)

Our quasi-Kantian principle disqualifies every precept that has the agent choosing to do x if he thinks some other person wants to do y, choosing to do not-x if he thinks the other wants to do not-y, choosing to y if he thinks the other wants to not-x and choosing to not-y if he thinks the other wants to x. Thus suppose that both Adam and Eve choose to go to the movies both today and tomorrow, each thinking that the other wants to go tomorrow only. If the function holds for both, this is in disequilibrium. Were each to see where the other stood, each would change his mind and plan not to go tomorrow. This too would be in disequilibrium. If each learned about the other, each would now decide to stay home both days – and again there would be a disequilibrium, etc. Other disqualified precepts in this vein can be constructed.

One of the standard criticisms of Kant is that his principle does not take us far. The same might with justice be argued here. Only the most obviously perverse precepts are ruled out. Is there any way of going further? Note that in addition to general principles there may also be local ones. Each precept prescribes some pattern of social choosing to every person linked to the rest as set out in its antecedent. If a precept qualifies as *moral*, it may perhaps express an ethics of parenthood. Or it may express an ethics of friendship, or of neighborliness, or of professionalism, or of the old-school tie or whatever. The question we might ask is: when is a precept with this or that antecedent a *moral* precept? Given that the antecedent is such or so, what must the consequent choice-function be like in order for the precept to pass? What are the local principles a precept must satisfy over and above the general principles that must be satisfied in every case?

In one special context, this question has been much discussed. The antecedent of a precept expressing an ethics of political office reports that the agent has been appointed to decide on collective actions for the group. What principles apply here? That is, what conditions are properly set on the consequent choice-function? The discussions grow out of an awareness that the initially likely conditions do not all hang together, that any choice function that satisfies some of them must fail to satisfy others. This is the upshot of Arrow's analysis, which (recast in choice-functional terms) establishes what can be read as a set of theorems of meta-ethics.[4]

Some of the most recent difficulties can be resolved in our terms here. Recall that an agent's residual set is a function of the others' interests *and* of the issue the agent faces. The Arrovian principles of political ethics all are issue-independent. They are meant to apply whatever the issue involved. Hence the paradox of the Paretian liberal.[5] We can avoid this if we hold that the Pareto principles and the liberal principles apply in distinct issue contexts. Where the issue is that of war and peace or of the regulation of the economy, Paretianism applies. Where the issue is that of what religion people will follow or of what they will read, liberalism applies, though, again, issue-relativity provides for different versions in different sub-contexts. Where the issue is whether *everyone* will be obliged to be a Catholic or whether *everyone* will be obliged to read the *New York Times,* each person might be allowed to be semi-decisive over the affirmative – each might be allowed a veto. Where the issue is whether *A* alone will be obliged to be a Catholic, *A* might be allowed to be fully decisive – allowed to dictate what goes into the residual set. At any rate, Paretianism and liberalism cannot come into conflict. The problem is that of specifying the private issues. A choice function may have to do this by enumeration. (The American Constitution reads, 'Congress shall make no law respecting the establishment of religion, or prohibiting the free exercise thereof; or abridging the freedom of speech, or of the press; or . . . .')

But let us move away from political ethics. Are there principles for precepts in other fields? We must first look at the concept of interests, to which we have paid no attention as yet. We can adapt some old ideas to our purposes here. Let us say that a person's *actual* interests need not be the same as his *true* interests. Also that each person has interests that are true in the other-regarding respect, and actual otherwise. Also that everyone has interests that are true in certain other respects only. These distinctions will be taken to mark out conditional value-structures, the conditions in each case being a set of beliefs, possibly about some other people, but possibly also not. Thus a person's actual interests are his value structures conditioned on what in fact he believes, his true interests are those he would have if he believed the truth only, his interests true in other-regarding respects those he would have if he believed the truth about others and otherwise believed what in fact be believes, etc. Think of a person's interests as multidimensional, each dimension reflecting his values in some context of possible beliefs. A person's values may or may not be the same in these different dimensions.

How many dimensions we note depends on various factors. But it might be held that, wherever we are obliged to defer to others, we ought to attend to each of the dimensions we have discriminated. This might, for a start, be expressed as a condition on the consequent choice-functions of those precepts that direct us to defer to some specific person *O*. Call this condition the condition of *tolerance*. It says that, whatever the issue, the choice function must identify a residual set for every possible combination of *O*'s value structures in the relevant belief contexts. We might go on to require that the choice function be *univocity respecting* – that it be such that, whatever the issue, where *O* prefers one proposition to a second in each of these belief contexts, the residual set does not contain the second proposition. The choice function might also be required to be *option focused* – to be such that, whatever the issue, the residual set is fully determined by *O*'s valuations in these contexts of the options involved. Finally, the choice function might be required to be *unselective* – to be such that, in no issue, do *O*'s valuations in one of these contexts alone determine the residual set.

These last two conditions may well be too strong. They may have to be weakened so as to refer to issues of specific sorts only. But let us keep to those issues for which all four conditions might be intended. We can see at once that they do not all hang together – that any choice function that satisfies some of them must fail to satisfy others – or rather that they do not all hang together if our choices are required to be consistent in certain respects when *O*'s interests are consistent. The four conditions correspond exactly to the Arrovian conditions of unrestricted domain, Pareto-optimality, independence of irrelevant alternatives and nondictatorship – the consistency requirement corresponding to the Arrovian requirement of social rationality. The basic impossibility proofs carry over. They establish that we cannot, in any issue in which we defer to some single person, always be tolerant, univocity respecting, option focused and unselective if we want our choices to be consistent when this person's interests are.

This bears only on those cases in which we defer to some single other person *O*. The four fundamental conditions can be extended. If this is done, an analogous conclusion holds for those cases in which we defer to *O* and *N* and *M* . . . collectively.

Here is more meta-ethics. What does it all show? One message is that there still is room for discussion on what a moral precept is like, and so on what makes up an ethics. We are not done with philosophy. I think it shows too – and this was the purpose – that the analysis of sociality can change our philosophical perspective. We are now thinking of moral constraints in terms very far from the usual, and we are facing questions we could not have raised before. Where this will get us remains to be seen.

Rutgers University

NOTES

[1] For a fully worked-out analysis of the evidence along these lines, see J. S. Brew, 'An Altruism Parameter for Prisoner's Dilemma', *Journal of Conflict Resolution* **17** (1973), 351–67.

[2] See my 'Beyond Utilitarianism', *Journal of Philosophy* **68** (1971), 657–66, and 'Democracy and Interdependent Preferences', *Theory and Decision* **3** (1972), 55–75.

[3] I argue the point in a forthcoming book.

[4] For a related reading of Arrow, and of political ethics in general, see Charles R. Plott, 'Ethics, Social Choice and the Theory of Economic Policy', *Journal of Mathematical Sociology* **2** (1972), 181–208.

[5] The paradox of the Paretian liberal is Amartya Sen's. See his *Collective Choice and Social Welfare,* Chapter 6*, Holden-Day, San Francisco, 1970. Sen surveys the literature in his 'Liberty, Unanimity and Rights', mimeographed.

ALAN R. TEMPLETON AND EDWARD D. ROTHMAN

EVOLUTION AND FINE-GRAINED ENVIRONMENTAL RUNS*

1. Introduction

The world in which we live is highly heterogeneous in both space and time. Organisms must live and reproduce in the face of possible changes in both their physical environment (e.g., weather) and their biotic environment (e.g., interactions with other organisms). Much of this immense array of environmental diversity is not controllable or even predictable by the individual organism. In such cases the exact array of possible environmental states an organism experiences throughout its lifetime must be treated as a random variable.

Another level of heterogeneity emerges when dealing with a population of organisms rather than just an individual. Recent studies in population genetics have demonstrated that there is extreme diversity between individuals of the same species with respect to the hereditary information contained in their genes (Lewontin, 1974). Genes are defined to be the basic units of hereditary information and molecularly correspond to a linear sequence of chemical subunits called nucleotides in a DNA molecule. It takes thousands of different types of genes (called gene 'loci') to code for all the information needed during the development and functioning of a complex organism. Genes at different loci almost always differ in their nucleotide sequence since they usually code for basically different types of information. However, this is not the type of genetic diversity referred to above. The genetic diversity of interest to evolutionary biologists rests on the fact that DNA can be replicated and passed on to future generations. Hence, a given type of gene has an existence beyond the individual since it can through time come to exist simultaneously in several different individuals. At the population level we can group together all those genes that occupy homologous loci irrespective of the particular individual which bears them. However, the replication of DNA is not always 100% accurate; mistakes called 'mutations' occasionally occur. Consequently, genes occupying the same locus do not necessarily have identical nucleotide sequences. These alternative forms of homologous genes are called 'alleles'. The genetic diversity referred to above is in fact 'allelic'

Hooker, Leach, and McClennen (eds.), Foundations and Applications of Decision Theory, Vol. II, 131–183. *All Rights Reserved*

diversity. This allelic diversity can also exist at the individual level in most sexually reproducing populations. Such populations are commonly 'diploid', a word which refers to the fact that an individual usually receives two genes for each locus; one from the father and one from the mother. If an individual inherits an allele from his father that is different from the one from his mother, this individual is said to be 'heterozygous'. If both genes are identical in allelic state, the individual is said to be 'homozygous'.

To make these concepts concrete, consider a particular locus in man, the homoglobin β-chain locus. This is a locus consisting of 438 nucleotides which stores the information needed to produce one of the protein components (the β polypeptide) of hemoglobin, a molecule that transports oxygen to the cells. A 1973 survey revealed 106 known allelic forms at this locus in man, each represented by a unique nucleotide sequence in the DNA (Hunt and Dayhoff, 1974). Two of the more common alleles at this locus are the '*A*' allele and the '*S*' allele. They differ at only one nucleotide out of the 438, yet the *S* allele when homozygous causes the often lethal disease sickle-cell anemia, while homzygous *AA* or heterozygous *AS* individuals do not suffer from such an anemia. However, if an *AS* individual is exposed to the environmental state of malarial infection, his resistance to malaria is much greater than that of an *AA* individual. This illustrates a very important point: that individuals of the same species can differ in their response to an environmental state because they inherited different alleles. Thus, at the population level, we see yet another level of diversity when dealing with the biological impact of random environments; namely, that different individuals within the population can respond differentially even to identical realizations of the environmental process if they have different genotypes.

Because such responses represent a gene–by–environment interaction, geneticists have been very careful in distinguishing between the concepts of an individual's 'genotype' – the kinds of genes he inherits from his parents – and his 'phenotype' – his actual manifested physical and behavioral characteristics. One does *not* inherit a phenotype, but rather a genotype which influences the phenotypic responses (both physical and behavioral) an individual makes when subjected to a given array of environments. If this array of environments is in some sense random, then the phenotype of an individual must also be regarded as a random variable. Thus, when modeling a population inhabiting a random environment one must take into account the fact that (1) an individual's phenotype is often a random variable and (2) the probability distribution describing the phenotypic response can vary between individuals of the same population due to genotypic variability.

Of the possible phenotypic measures one can make on an individual, the one that is of paramount importance to the evolutionary biologist is 'fitness'. Fitness refers to how many copies of an individual's genes are passed on to the next generation. This in turn depends upon the organism's ability to survive and reproduce. Fitness is influenced by the expression of the individual's genes in response to the array of environments experienced throughout the entire lifespan until reproduction ceases. Fitness, like any other phenotype, is a genotype-by-environment interaction, and hence must be treated as a random variable if interacting with a random environment. Alleles that tend to be associated with high fitness phenotypes will increase in the populations because more copies of that allele are passed on, while alleles that are associated primarily with low fitness phenotypes decrease in frequency. This is the basis for natural selection, and it can result in changes over time of the frequencies with which certain alleles are found in the population. Changes in the frequencies of certain alleles or complexes of alleles shared by a population over time are by definition evolution. Forces other than natural selection can cause evolutionary change, but in this paper we will confine ourselves solely to evolution caused by natural selection.

We are now in the position to state the problem of evolution in a random environment. This problem can be stated in two different ways. First, if fitness is influenced by a random environment, how does this affect the evolution of a population; i.e., how does fitness being a random variable alter the genetic composition of a population over time? Second, given that an organism or a population of organisms is faced with an uncertain environment, what is the optimal phenotypic response of the organisms or population to this uncertainty? These two statements of the problem of evolution in heterogeneous environments have sometimes been viewed as equivalent, but we feel they are expressions of two very distinct problems. In fact, the second expression is basically not even evolutionary, although it is usually tied in to evolutionary theory by assuming the 'optimal' population corresponds to the population that will actually evolve. However, it is not clear at this time what is optimized during evolution, if anything.

Richard Levins, beginning in a paper written in 1962 and culminating in a monograph (1968), tried to answer the second expression of the problem of evolution in changing environments. There are three basic components to Levins' model. First, there is a model of environmental heterogeneity. Levins dealt primarily with discrete environmental states such that one and only one could be experienced by an individual at a given time. This environmental heterogeneity is assumed to be completely described by the average frequencies with which these environments are encountered. However, to evaluate the biological impact of these random environments, it is further necessary to relate the time scale on which the environmental states are realized to the time scale of the individual's life-span. In particular, Levins noted two possible extremes. On one extreme, the life-span of the organism is long in relation to the time scale on which environmental transitions occur. Thus, an individual organism experiences his environment as a succession of possibly different states within its own life-time. Furthermore, the life-span is assumed to be sufficiently long that the frequency with which an individual experiences a given environmental state within its own life-time is in fact the average state frequency which is assumed to be the same for all generations. The environment is therefore uncertain for the individual, but the average or expected environmental conditions are identical for all individuals and all generations. This type of environmental heterogeneity is said to be 'fine-grained'. The other extreme is the 'coarse-grained' environment. Under this model, any particular individual lives out the entire selectively relevant portion of its life-span under a single environmental regime. However, different individuals may live under a different environmental state. This could occur because the environment is broken into spatial 'patches' with the patches sufficiently large that an individual has essentially no chance of leaving the patch during his lifetime. It can also occur when the environment is constant for any given generation, but fluctuations occur between generations. Coarse-grained environments are thus constant for the individual, but random for the population over space and/or time.

The second component of Levins' model is the fitness response of the individual (or population) to the environmental heterogeneity. Corresponding to each discrete environmental state, Levins assigns a discrete fitness response. Thus, if there were n environments, the fitness response of a population which is monomorphic (i.e., all individuals have the same type of response) could be described by the set $(w_1, w_2, \ldots, w_n)$ where w_1 is the fitness response to the *i*th environmental state. One can incorporate this information into a 'fitness set' by creating an n dimensional hyperspace with each axis corresponding to the possible fitness responses in a specific environment. The set $(w_1, \ldots, w_n)$ can then be represented as a point in this n-dimensional hyperspace. The set of all such points for the different monomorphic populations constitutes the fitness set. Finally, monomorphic populations can be mixed to yield polymorphic populations. The fitness set is therefore extended to allow polymorphic populations and this 'extended fitness set' is simply the convex hull of the fitness set.

As Levins (1968) points out, the fitness set does not uniquely determine the fitness of the population, and it is necessary to define an 'adaptive function' which relates the fitness response to any given environment and the frequency of that environment to an overall fitness measure. The optimal population is defined as that population which maximizes the value of the adaptive function. In essence, the adaptive function defines what one means by 'optimal'. The adaptive function also depends upon the grain of the environment. For example, Levins used the arithmetic mean of the fitness responses weighted by the environmental state distribution as the adaptive function for fine-grained environments, and he used the geometric mean for coarse-grained environments. The form of the adaptive function is critical to Levins' formulations, and the definition of 'optimal' implicit in this adaptive function is the third major component of fitness set models.

There are many conceptual similarities between Levins' fitness set approach to evolution and a decision theoretic approach to no-data decision problems. In no-data decision problems, n 'states of nature' exist which correspond to the n environments. A 'prior distribution' is defined over these states of nature analogous to frequencies of environments in the biological situation. The decision theorist has available to him several actions or decision rules that may be either 'non-randomized' decision rules that correspond to monomorphic populations or 'randomized' decision rules that correspond to polymorphic populations. A loss function is defined that associates a loss for each nonrandomized rule in each of the n states of nature. The loss associated with a mixture of nonrandomized rules is the mixture of losses. Losses of all possible rules or actions thus form a convex set which is the convex hull of the loss set for the nonrandomized rules. The object is to find a rule which in some way results in a 'small' loss. In the biological situation instead of loss there is fitness and 'high' fitness is in some way desirable. Because of these similarities, we were able to apply many theorems and concepts developed in decision theory to the problem of fitness set analysis (Templeton and Rothman, 1974). In particular, we showed that the conclusions based on fitness set analysis – even in their qualitative aspects – are very sensitive to the precise definition of 'optimal' that is used.

One serious omission in much of Levins' work and in our own previous paper is the lack of genetic constraints. This is indeed a serious omission because evolution is basically a genetic process. Without genetic constraints we are unable to answer the first and truly evolutionary expression of the problem of random environments; namely, if fitness is influenced by a random environment, how does this affect the genetic composition (evolution) of a population? It is this question that is our primary concern in this paper, and not the second non-evolutionary expression of the problem. We must therefore add a fourth component to Levins' model – genetic constraints.

Levins, himself, did incorporate genetic constraints into some of his models (Levins and MacArthur, 1966; Levins, 1968), but did so only in a limited manner. He realized that the genetic constraints would eliminate certain types of populations from ever being realized. For example, the 'optimal' human population in a malarial region without access to anti-malarial drugs would be a population consisting solely of *AS* heterozygotes at the hemoglobin β-chain locus since they do not get anemia nor serious cases of malaria. However, the laws of Mendelian genetics insure that one-half of the offspring of an *AS* individual will be homozygous, so that it is impossible to achieve a population consisting solely of *AS* humans. Levins took such factors into account by redrawing the extended fitness set to restrict it to the genetically realizable populations. However, he retained all other features of his original model. In particular, he retained the notion of an adaptive function and thus imposed a definition of 'optimal' upon the system. However, the importance of genetic constraints is far greater than just the exclusion of certain types of populations. In fact, the genetic constraints completely undermine the philosophical basis of fitness set analysis.

As pointed out earlier and as discussed in more detail in Templeton and Rothman (1974), fitness set analysis can be regarded as a rewording of no-data decision theory. However, as soon as genetic constraints are introduced into this model, this is no longer a 'no-data decision' problem. Remember, evolution occurs as the frequency of various genes or gene complexes is altered in a population over time. Moreover, natural selection provides a mechanism by which gene frequencies are influenced by the state of past and present environments. Consequently, information about the environment is contained within the gene frequencies. This is particularly true for fine-grained environments in which the environmental process is assumed to be stationary over time (Lewontin, 1966). Considerable information about the environment can be built up and stored in the form of an array of gene or gene-complex frequencies. This information invalidates the use of theories appropriate for no-data decision problems in studying evolution in random environments.

As an example of the role of genetic constraints, consider the special case in which a random environment interacts with the genotypes determined by a single locus with two alleles in a discrete-generation, sexually reproducing diploid population. Let the two alleles be symbolized by **A** and **a**. There are three possible genotypes: **AA**, **Aa** and **aa**. Let p_t = the allele frequency of **A** among the pooled zygotes at generation t. Under the assumption that the population size is effectively infinite (i.e., we will ignore sampling error of alleles as a source of evolutionary change) and that individuals mate at random with respect to this locus, the genotype frequencies of the zygotes at generation t are:

Genotype	AA	Aa	aa
Zygotic Frequency:	p_t^2	$2p_tq_t$	q_t^2

where $q_t = 1 - p_t$ = the zygotic allele frequency of **a** at t. (This is the 'Hardy–Weinberg Law'; for a proof, see any introductory genetics text.) If $(\sigma_t, \rho_t, \tau_t)$ are the realized fitnesses of the genotypes **AA**, **Aa** and **aa** respectively at generation t, then the allele frequency of **A** among the zygotes at generation $t + 1$ is

$$(1\text{-}1) \qquad p_{t+1} = \frac{\sigma_t p_t^2 + \rho_t p_t q_t}{\sigma_t o_t^2 + 2\rho_t p_t q_t + \tau_t q_t^2}.$$

Note that (1-1) completely defines the evolutionary change (i.e., allele frequency change) that occurs between generation t and $t + 1$. Thus, given a sequence of realized fitnesses it is possible to use (1-1) recursively to completely define the evolutionary process. As shown by Karlin and Lieberman (1974), this is also true in a great many cases when the fitnesses are treated as random variables rather than treated as realized constants. There is no need to invoke *any* outside definition of optimum in order to study the evolutionary process. This means that when the three components of Levins' basic model are coupled with genetic constraints, the model is overspecified. Hence, one component can always be allowed to vary as a function of the other three.

The fact of critical importance contained in the above paragraph is, that when evolution is *modeled* as a genetic process (and we feel that it should be, since all modern texts dealing with evolution *define* it as a genetic process), it is unnecessary to invoke optimality criterion. This simply reflects the scientific belief that evolution is a genetic process governed by natural laws and constraints, and requires no extrinsic directive force. Yet many of the theoretical evolutionary models have used such extrinsic optimality criteria (e.g., Levins, 1968; Templeton and Rothman, 1974). If we examine in more detail the theoretical literature on evolution of this century, we can distinguish two major currents. First, there is what we will call the 'population genetic' literature which was founded by the fundamental works of Fisher (1930), Haldane (1932), and Wright (1932). The models from this literature generally dealt with the genetic mechanism of evolution given population genetic constraints and known fitness patterns. The environmental input into the evolutionary process was only implicit in the fitness model, but usually not explicitly stated. The dynamics and equilibrium of the genetic composition of the population were deduced from the fitness process and the population constraints. Optimality criteria are generally *not* found in this literature, and when they do appear they are deduced from the population genetic constraints (e.g., Fisher's fundamental theorem of natural selection, 1930), and not assumed *a priori*.

The second major current is what we shall call the 'ecological adaptation' literature. These are works mainly by non-geneticists who are primarily concerned with the adaptive results of evolution rather than the genetic mechanism by which they are achieved. The environment is necessarily an important component of these models, and the main purpose is to discover which of the morphological or behavioral strategies available to the organism is the 'best' adapted to the environments it encounters. The focus is upon the individual and populational constraints are ignored; hence, these models are not truly evolutionary since only populations can evolve, not individuals. To bypass this non-evolutionary aspect of these models, optimality criteria are assumed *a priori* to identify the 'best' phenotype. It is this literature in which one encounters such phrases as 'optimal clutch size', 'prudent predators', 'optimal foraging strategies', 'adaptive function', etc. We have already illustrated the difference in attitude between these two major current by our two statements of the problem of evolution in heterogeneous environments given earlier in this introduction. The first version is that of a 'population geneticist', the second of an 'ecologist'. It is in the ecological literature that one finds many applications of decision theory to biology, but such applications are rarely found in the population genetic literature.

There is beginning to be a fusion of these two areas of modern evolutionary theory to form a third area – theoretical ecological genetics. In this area, the theoretician is interested both in the genotypic evolution of the population as well as its phenotypic adaptations to an explicitly stated environment. There are two basic approaches such a theoretician would take. First, he could start with the theoretical framework developed in the ecological literature that makes frequent use of decision theory, and overlay upon it genetic constraints to exclude certain types of populations (e.g., Levins and MacArthur, 1966). However, as we have already pointed out, such a model is overspecified and thus makes inefficient use of the available information. More importantly, we have previously emphasized that the introduction of genetic constraints undermines the philosophical basis of using extrinsic optimality criteria. Therefore, we suggest that a much stronger approach would be to modify the population genetics models by explicitly developing an environmental model rather than simply invoking a ready-made set of fitnesses. The tools of population genetics – rather than decision theory – could then be used to predict both the population's 'adaptation' to the environment (i.e., which morphological and behavioral strategies will actually evolve in response to a specified environment) and the genetic process by which this adaptation was achieved. Thus, the 'adaptive process' is examined in its original, evolutionary sense as expressed by Darwin, and is not necessarily viewed as an 'optimizing process'. The purpose of this paper is to illustrate this type of model building with the problem of evolution in heterogeneous environments.

As should be evident by now, our feelings about the applicability of decision theory to evolutionary problems are essentially negative. Decision theory has been applied in the past to predict the adaptive end results of evolution, but, as we shall illustrate in this paper, the predictions from these models often fail to be confirmed by the explicitly evolutionary models of population genetics. Because of this, we feel that in the future decision theory will play an ever decreasing role in evolutionary biology. By this we do not mean to imply that the population genetic constraints, together with a specification of the environment and fitness responses, entirely renders the notion of individual morphological or behavioral strategies irrelevant to biological evolution. Rather, we are saying that such a model already builds in a description of the strategic choices. In this sense, the future may reveal that evolutionary biology will be increasingly applied to decision theory rather than the other way around.

2. Equilibrium Populations in a Heterogeneous, Fine-Grained Environment

As mentioned in the Introduction, there are four major components to the types of evolutionary models dealing with heterogeneous environments: (i) the nature of the environmental heterogeneity, (ii) the fitness response to this heterogeneity, (iii) the genetic constraints operating upon the evolving population, and (iv) the definition of the 'optimal' population or response. In this Section we make assumptions about (i), (ii) and (iii), but not (iv). Rather, we use some techniques devised by Karlin and Lieberman (1974) to specify the nature of the stable equilibrium population that arises from our assumptions about the other components. This allows us to investigate our first expression of the problem of evolution in random environments. To see if these results can be related to the second expression of this problem that Levins was concerned with, we then investigate whether these stable equilibrium populations optimize some 'adaptive function' or not.

Although we use the stability criteria of Karlin and Lieberman, our approach for studying evolution in heterogeneous environments differs considerably from theirs. They incorporated environmental heterogeneity into their model by making fitness a random variable directly without relating it to some underlying model of environmental variability. This approach obscures the role of environmental heterogeneity since all parameters of the model refer to fitness rather than environments. To avoid this problem, we first develop a stochastic model of environmental heterogeneity. We then define a deterministic fitness response function to a particular array of realized environments. By regarding the probability distribution over environmental arrays that is generated by our environmental stochastic process we can secondarily treat fitness as a random variable. The strength of this procedure is that the fitness process is explicitly described by the parameters determining the environmental stochastic process and those defining the organism's response to the environments it encounters. Consequently, we can deal directly with the evolutionary impact of different types of environmental heterogeneity and/or different types of responses to these environments. Treating fitness itself as the primary random variable causes these aspects of the process to be subsumed into a black box.

We begin with a stochastic model of environmental heterogeneity. The model we use in this paper has two discrete environmental states, say '0' and '1'. During any particular time unit, the environment is either in one state or the other. Trasitions between time units are governed by the stochastic matrix:

$$\begin{array}{cc} & \begin{array}{cc} 0 & \quad 1 \end{array} \\ \begin{array}{c} 0 \\ 1 \end{array} & \begin{pmatrix} 1-\alpha & \alpha \\ \beta & 1-\beta \end{pmatrix} \end{array} \tag{2-1}$$

where $1-\alpha$ is the probability of being in state 0 given the process was in state

0 during the previous time unit, α is the probability of being in state 1 given 0 previously, β is the probability of being in state 0 given 1 previously, and $1-\beta$ is the probability of being in state 1 given 1 previously. The environmental process is thus assumed to be Markovian as governed by the above transition matrix. Such a model has been used to successfully describe rainfall (Gabriel and Neumann, 1962). Despite its simplicity, this model can describe a wide range of conditions. The average frequency of environments 0 and 1 over a large period of time are $p_o = \beta/(\alpha + \beta)$ and $p_1 = \alpha/(\alpha + \beta)$ respectively. By choosing α and β appropriately, we can describe environments differing in the average frequency of their states. More importantly, an infinite number of α's and β's can yield identical p_0's and p_1's. Thus, we can contrast environments with the same average state frequencies, but which differ in how transitions are made between states – e.g., we can have 'runny' environments with long sequences of one state or the other very probable, or non-runny environments where rapid alternation of states is most likely. The ability to describe 'runs' is an important feature of this model since much of the environmental heterogeneity that is known to have great biological impact is defined in terms of runs. For example, droughts are known to greatly affect many plant and animal species, yet a drought only occurs when a run of 'dry-days' is experienced. Simply specifying the average frequency of dry days does not in general define the chances of encountering a drought. Hence, some measure of the 'runniness' of the environment is often a critical parameter for biological problems.

We now discuss how the organism experiences and responds to this environmental heterogeneity. Suppose the organism of interest samples a set of L consecutive time units from a stationary process governed by (2-1) where L represents the portion of the lifespan that is subjected to the environmental process being modeled. We assume that L is sufficiently large that the organism experiences within the timespan of L the environmental states 0 and 1 with frequencies p_0 and p_1 respectively; that is, L is large relative to the time scale of environmental transitions. This assumption is equivalent to saying the environment is 'fine-grained' (Templeton and Rothman, 1974). However, some of the extensions of our basic model include some aspects of 'coarse-grained' environments in them. We further assume the environments which each generation encounters are governed by identical probabilistic rules, but that all individuals of a given generation share a single realization of the random environmental process (i.e., there is no independent sampling within a generation), although this restriction will be somewhat relaxed in certain sections. An example of such an environmental model would be the states of weather that affect a local population of animals or plants inhabiting a restricted geographical area. All individuals in such a population at a given time would experience an identical array of states of the weather, but different generations would experience different realizations.

We next assume each individual has some inherent fitness components C which is a function only of its genotype and unaffected by environmental state. A second component of fitness that is multiplied by C to yield total fitness is affected by the environmental heterogeneity encountered. In this paper we will limit our discussion to the case where this second fitness component is unaffected by state 0, but can be lowered by experiencing state 1. We further assume the amount of this lowering of fitness depends upon how long the organism has been continuously exposed to state 1. In other words, the fitness effects are a function of the run length of state 1, where a run of 1's of length n is defined as a state 0 followed by n consecutive 1's followed by another 0. Hence, the overall fitness of the organism is assumed to be of the form:

$$W = C\prod_i w_{n_i} \tag{2-2}$$

where i indices all runs of 1's experienced during L and w_{n_i} is the conditional fitness effect associated with run i given it has a length of n_i. This is a realistic fitness function for a great many biological situations, such as drought or famine, since it allows the fitness impact of the environmental state to be a function of the environmental context in which it occurs.

We must now couple (2-1) with (2-2) in order to treat fitness as a random variable. In using many of the criteria of Karlin and Leiberman (1974), it is only necessary to have the first two moments of the log fitness:

$$\log W = \log C + \sum_i \log w_{n_i}. \tag{2-3}$$

We use the results given in Good (1973) on the probability generating function for run lengths in a Markov chain to obtain:

$$E(\log W) = \frac{\alpha \cdot \beta}{\alpha+\beta}\,(S_1\,[\beta(L+1) - p_1]\; + \beta S_2) + \log C$$

$$\mathrm{Var}\,(\log W) + Lp_1\beta^2\,[S_3 - p_1\,\beta^2 S_1(2S_2 + S_1) + 2p_1^2\beta S_1^2] \tag{2-4}$$

where

$$S_1 = \sum_{n=1}^{\infty} (1-\beta)^{n-1} \log w_n$$

$$S_2 = \sum_{n=1}^{\infty} n(1-\beta)^{n-1} \log w_n = -\frac{\partial}{\partial\beta} (1-\beta)S_1$$

$$S_3 = \sum_{n=1}^{\infty} (1-\beta)^{n-1} (\log w_n)^2 .$$

Equations (2-4) are obtained under the assumption the sequence begins in state 0, but we will evaluate (2-4) only under the assumption of large L so that this initial effect will not influence the results (Good, 1973). With large L and the type of fitness functions we use later;

$$(2\text{-}5) \qquad E(\log W) = \frac{\alpha\beta^2}{\alpha+\beta} S_1 L + \log C$$

and Var (log W) is unchanged.

We now consider a special case of (2-2) in which

$$(2\text{-}6) \qquad w_n = \begin{cases} 1 & \text{if } n \leqslant d \\ e^{-\lambda(n-d)} & \text{if } n > d \end{cases}$$

Under this model the organism can withstand a run of 1's up to length d with no ill effects, but thereafter fitness declines exponentially at rate λ. In other words, only prolonged exposure to state 1 causes damage to the organism. This model reflects the biological fact that most organisms can effectively buffer themselves against even severe environmental perturbation if this perturbation is not of great duration. With this model and logs to the base e.

$$(2\text{-}7) \qquad \begin{aligned} S_1 &= \lambda(1-\beta)^d/\beta^2 \\ S_2 &= S_1(d-1+2/\beta) \\ S_3 &= \lambda^2(1-\beta)^d(2/\beta-1)/\beta^2 . \end{aligned}$$

Hence, from (2-4) and (2-5),

$$(2\text{-}8) \qquad \begin{aligned} E(\log W) &= \log C - \lambda L p_1(1-\beta)^d \\ \text{Var}(\log W) &= L\,\lambda^2 L p_1(1-\beta)^d \left[2/\beta - 1 + p_1(1-\beta)^d\left(\frac{2p_1-4}{\beta} - 2d+1\right)\right] . \end{aligned}$$

We now introduce genetic constraints into our model and explore the types of equilibrium populations that evolve. In this section we are primarily interested in investigating in what sense the equilibrium population is 'optimal'.

Consequently, we will only consider a simple genetic model that allows us to explore this problem, although more complex genetic models will be used in later sections. For now, we confine ourselves to a one-locus, two-allele diploid model with complete dominance for fitness effects such that

Genotype	**AA**	**Aa**	**aa**
Fitness	1	1	W

where W is of the form described by (2-2) and (2-6). In this model the **AA** and **Aa** genotypes have a homeostatic phenotype that is insensitive to run's of 1's, but the **aa** homozygote suffers from runs of 1 longer than d. Since in biology as in other fields there is no such thing as a free lunch, we assume the dominant phenotype achieves its homoeostasis at a price, as reflected by $C > 1$. Thus, if no runs of 1 longer than d were every encountered, **aa** would have an unambiguous fitness superiority to the dominant phenotype. A possible example of such a situation is given by Schwartz (1969). In maize, the enzyme alcohol dehydrogenase occurs in the kernel and is active until shortly after germination. Schwartz isolated a mutant at the Adh_1 locus that when homozygous fails to produce any active enzyme. This allele is given the symbol $Adh_1{}^0$, and the active allele the symbol Adh_1. With respect to *Adh* activity, Adh_1 is a dominant allele. Under aerobic germination conditions, the $Adh_1{}^0/Adh_1{}^0$ homozygous kernels do just as well as $Adh_1/$- kernels, but *Adh* is known to be important in energy production in maize under anaerobic conditions. In nature, anaerobic conditions would result from an excessive amount of water in the soil following a heavy rainfall. Under flooding conditions, the recessive *Adh*-null kernels no longer did well. If subjected to 72 hours of continuous soaking, all of them failed to germinate. $Adh_1/$- kernels, however, were insensitive to this soaking. This is the same type of fitness pattern described above, with Adh_1 corresponding to '**A**' and $Adh_1{}^0$ to '**a**'.

As mentioned in the introduction, an evolutionary process is genetically defined by the sequence of allele frequencies among zygotes over generations. Let q_j be the frequency of allele '**a**' among zygotes at generation j and $W(j)$ the realized fitness for generation j. Assuming random mating at this locus (random union of the two types of gametes, '**A**' and '**a**') and discrete generations, the allele frequency at generation $j+1$ is given by:

$$(2\text{-}9) \qquad p_{j+1} = [W(j)q_j^2 + q_j(1-q_j)]/[(1-q_j)^2 + 2q_j(1-q_j) + W(j)q_j^2]$$

Equation (2-9) essentially defines the evolutionary process consistent with our genetic constraints. This process reaches equilibrium when the allele frequency becomes stabilized. Since W is a random variable, we say that q^* is a stochastically locally stable equilibrium if for any $\epsilon > 0$, there exists a $\delta(\epsilon) > 0$ such that whenever $|q - q^*| < \delta(\epsilon)$, for an initial frequency q, then Probability $(1\,\text{im}_{j\to\infty}\, q_j = q^*) > 1 - \epsilon$ (Karlin and Lieberman, 1974). Karlin and Lieberman have shown that fixation of the **a** allele ($q^* = 1$) is stochastically locally stable if

$$E(\log W) > 0$$

or, when the fitnesses are of the form described by (2-6),

$$\text{(2-10)} \quad \log C > \lambda L p_1 (1-\beta)^d .$$

From (2-10) we can see that the **a** allele tends to become fixed as C increases (as the fitness advantage increases of **aa** over **A**– in the 'favorable' environment with no long runs of 1's), as λ decreases, or d increases (as the sensitivity of **aa** to runs of 1's decreases), as the frequency of state 1 (p_1) decreases, as the 'runniness' of the environment $(1-\beta)$ decreases, or as the lifespan (L) decreases. This last observation is interesting since it means that in long-lived organisms, the types of fitness effects we are now modeling here become increasingly important and selection tends to favor more and more those genotypes with the greatest homeostatic capabilities even though they may have considerably lower fitness under the environmental states most probably encountered.

The condition for fixation of the **A** allele ($q^* = 0$) is (Karlin and Lieberman, 1974)

$$\text{(2-11)} \quad E(W) < 1.$$

If W is close to 1 (i.e., n W close to 0), we can put the above inequality in terms of log W with the expansion

$$E(W) = E(e^{\text{n} W}) = 1 + E(\text{n}W) + \frac{\text{Var}(\text{n}W)}{2} + \frac{(\text{E}(\text{n}W))^2}{2}$$

But

$$\frac{[E(\text{n}W)]^2}{2} < |E(\text{n}W)| \sim 0.$$

Hence, (2-11) is satisfied whenever

$$\text{(2-12)} \quad 0 < \text{Var}(\log W) < -2E(\log W).$$

The variance is required to be positive since (2-11) does not apply when the fitnesses are constants. When

$$\text{(2-13)} \quad \mathrm{Var}(\log W) > -2E(\log W) > 0$$

fixation for either allele almost never occurs. This results in a 'polymorphic' situation with both alleles present in the population, although some intermediate q^* does not necessarily exist that is stochastically stable.

To illustrate these results, consider the following specific examples. Let the matrix describing the environmental state process be

$$\text{(2-14)} \quad \begin{pmatrix} 5/8 & 3/8 \\ 3/4 & 1/4 \end{pmatrix}.$$

With this matrix, $p_1 = 1/3, p_0 = 2/3$ and the average length of a run of 1's 1's = 1.33 time units. Letting $L = 100$, $C = 1.1$, $d = 5$ and $\lambda = 0.1, E(\log W) = 0.0918 > 0$. Consequently, we get fixation of the **a** allele and the equilibrium population consists solely of **aa** homozygotes.

Now consider the environmental matrix,

$$\text{(2-15)} \quad \begin{pmatrix} 7/8 & 1/8 \\ 1/4 & 3/4 \end{pmatrix}$$

Once again $p_1 = 1/3$ and $p_0 = 2/3$, so both environments have the same 'average' state. However, the average run length of 1's is now 4, so (2-15) describes an environment that is much more 'runny' than that of (2-14). Using the same fitness parameters, $E(\log W) = -0.6957$ which means that fixation for **a** will almost never occur. Furthermore, (2-12) is satisfied so the equilibrium population consists solely of **AA** homozygotes. The contrast in the results obtained with (2-14) and (2-15) demonstrates the importance of 'runniness' in the environment. Also note that fixation of **A** occurs despite the fact the average run length is less than d. A biologist observing these organisms would thus conclude the environment is usually more favorable to **aa** individuals than to **A**-'s, yet the **aa**'s become extinct. The example of the *Adh* locus in maize studied by Schwartz (1969) seems to fall into this category. After Schwartz's initial studies under the 'normal' aerobic conditions, all his data led him to predict that the Adh_1 allele should be lost (i.e., the $Adh_1{}^0$ allele frequency should go to 1). Yet, almost all maize populations are fixed for Adh_1. Schwartz then did his soaking experiment to resolve this contradiction and stated 'that even though the gene [Adh_1] is not essential for normal development of the plant under the normal range of cultural conditions, it may be essential for survivial under adverse conditions'. This demonstrates the limitations of models that deal only with the 'expected' or 'normal' environment.

The impact of lifespan can be seen by letting $L = 3000$ for the case described by (2-14), all other parameters remaining the same. Now $E(\log W) = -0.0023$ and $\text{Var}(\log W) = 0.0162$. This satisfies (2-13) and a polymorphic population results. Thus, fixation of the **a** allele will not occur with $L = 3000$ although it did with $L = 100$. In both cases the 'average' environment is the same, and since d is greater than 1.33 this environment is usually more favorable to **aa** than to **A**- individuals. The impact of lifespan in this situation is particularly interesting because the definition of fine-grain is that the organism experiences the environment with its average frequency. This becomes more and more true as L increases, yet the impact of runs also increases in importance and causes the 'average environment' to become less meaningful. This example along with the previous one shows the danger of basing biological predictions upon statements about the 'average environment'.

To investigate the problem of optimality we now recast these examples in a fashion analogous to Levins' fitness sets. Levins defined the optimal population as the population that maximized an adaptive function which in turn was some increasing function of the fitness components. We will now consider the log fitness as a candidate for the 'adaptive function' as Levins himself did for fine-grained environments (Levins, 1962 and 1968, p. 18). The log fitness can be reformulated as

$$\log W = \log C - \lambda B$$

where B is the total number of time units spent in state 1 after at least d consecutive 1's had been previously experienced (such a time unit will hereafter be referred to as a 'bad' time unit because it lowers the organism's fitness). The expected value of B is easily calculated using the equations of Good (1973) by calculating the expected value of

$$\delta_n = \begin{cases} 0 & \text{if } n \leqslant d \\ n-d & \text{if } n > d \end{cases}$$

which yields $E(B) = Lp_1(1-\beta)^d$. This gives the same equation for $E(\log W)$ as before. The usual definition of optimum in fitness set analysis defines the optimal population as that one which maximizes fitness or some increasing function of it; i.e., the population with maximum

$$E(\log W) = \log C - \lambda E(B). \tag{2-16}$$

Equation (2-16) corresponds to an adaptive function in Levins' terminology of the form

$$A = w_o + w_1 E(B)$$

where $w_o = \log C$, a fitness measure of the organism's response to a 'good' environment (a run of 0's or a run of 1's less than or equal to d) and $w_1 = -\lambda$, a fitness measure for the 'bad' environment. The population with maximum fitness can be found graphically by the intersection of the extended fitness set with the lines defined by $A = w_o + w_1 E(B)$ as the w's change which has maximum A. The previously discussed examples are so graphed in Figure 1. The fitness set is found by graphing the fitness points $(w_o, w_1) = (\log C, -\lambda)$ for the two phenotypes, **A**- and **aa**. The point $(0.1, -0.1)$ corresponds to a population of **aa** homozygotes ($q^* = 1$), and the point (0, 0) correponds to a population of **AA** homozygotes ($q^* = 0$). The extended fitness set is obtained by drawing the line between these points. All intermediate points on this line correspond to polymorphic populations (an intermediate q). Given a value for $E(B)$, the maximal intersection with this extended fitness set and adaptive function lines can then be found. $E(B)$ is a function of the environmental matrix and the lifespan. For matrix (2-14) with $L = 100$, $E(B) = 0.035$, the intersection shown in Figure 1 occurs at the population consisting only of **aa** individuals ($q^* = 1$). This is also the equilibrium population that evolves with the given genetic constraints. With matrix (2-15) and $L = 100$, $E(B) = 8.82$ and the intersection occurs at $q^* = 0$; once again the same population that evolves under these conditions. In both these cases evolution seems to be maximizing average log fitness and our adaptive function is doing a good job of identifying the population that will actually evolve. Now consider the case when $L = 3000$ and the environment is described by (2-14). Now $E(B) = 0.98$ and the population with maximum log fitness is that with $q = 0$. However, our previous analysis showed this would not occur, rather the population will be polymorphic and lie on some intermediate point along the extended fitness set line. Consequently, the adaptive function is not maximized. This occurs because the genetic constraints that are operating on this system allow the higher moments of the adaptive function (in this case, its variance) to influence the evolutionary outcome. In this last example, the average log fitness value indicates that selection for the most part operates to eliminate the **a** allele. This is accomplished by eliminating **aa** homozygotes due to their average inferior fitness. However, the genetic constraints of random mating and diploidy insure that the zygotic frequency of these **aa** individuals is q^2. Consequently,

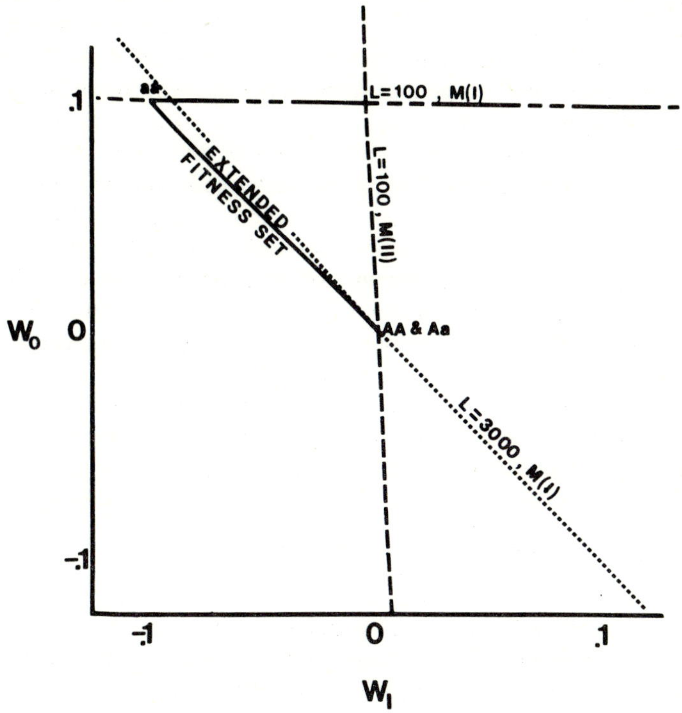

$$M(I) = \begin{matrix} & 0 & 1 \\ 0 & 5/8 & 3/8 \\ 1 & 3/4 & 1/4 \end{matrix} \qquad M(II) = \begin{matrix} & 0 & 1 \\ 0 & 7/8 & 1/8 \\ 1 & 1/4 & 3/4 \end{matrix}$$

Fig. 1. Fitness set graph of the example discussed in Section 2. The abscissa corresponds to the fitness effects of a 'bad' environment (a time unit in state '1' encountered after at least five previous, consecutive 1's), and the ordinate to a 'good' environment (all other time units). The **AA** and **Aa** (**A**-) individuals are plotted at the point (0, 0) since they are insensitive to this environmental heterogeneity. The **aa** individuals are at (—0.1, + 0.1) since they do better than the dominant **A**- phenotype in 'good' environments but worse in 'bad'. All populations that are polymorphic contain a mixture of **A**- and **aa** phenotypes and lie along the straight line connecting the points corresponding to **AA** and **aa**. This line is known as the extended fitness set and represents all possible populations that could exist. The broken and dotted lines are plots of the 'adaptive function' (average log fitness), which in this model are functions of L, the length of the lifespan affected by this environmental heterogeneity, and the environmental transition matrices, M(I) and M(II). The intersection of these adaptive functions with the extended fitness set identify the 'optimal' population for a given L and given environmental matrix. For further details, see the text.

as q decreases, q^2 decreases even faster and most of the **a** alleles are now found in **Aa** individuals. The genetic constraint of dominance protects these **a** alleles from selection since **Aa** individuals have a fitness of 1. All this means that as q becomes small, selection against the **a** allele becomes increasingly ineffective. Consequently, the rate of elimination of **a** is extremely slow. Furthermore, since $\mathrm{Var}(\log W) > -2E(\log W)$ in this case, there will be occasional generations during which **a** will be favored. Note that the genetic constraints operate to increase the effectiveness of selection favoring **a** as q increases. Consequently, this type of evolution proceeds at a faster rate than the elimination of **a**. The polymorphic situation evolves due to a balance between the variability in the direction of selection and the genetic constraints which influence the dynamics of the response to selection. The genetic constraints are therefore not minor nuisances, but are critical parameters in determining the outcome of evolution. Further examples of the impact of these constraints and how higher moments of a fitness function influence the equilibrium population will be given in subsequent sections of this paper.

3. The Evolution of Homeostasis

Homeostasis refers to the property of an organism to adjust itself to variable conditions, or to the self-regulatory mechanisms of the organism which permit it to stabilize itself in fluctuating inner and outer environments. It has long been recognized that the homeostatic properties of an organism have a genetic basis (Lerner, 1954; Lewontin, 1958; Thoday, 1955) and have therefore evolved. We can use the types of models discussed in the previous section to study the evolution of homeostasis since these models deal with an organism's response to various types of fluctuating environments. However, our emphasis will now be on 'optimal' fitness responses to environmental heterogeneity. In terms of the four major model components we will evaluate variations in the fitness response (ii) by some criterion of 'optimality' (iv) as a function of assumptions made about the nature of the environmental heterogeneity (i), and the genetic constraints (iii). We retain the same basic model for (i), develop a more general genetic model, and retain a fitness function of the form given by (2-2) and (2-6) but whose parameters are now free to vary. We need some definition of 'optimal' to evaluate the responses obtained as the fitness parameters vary. The results of the previous section and of a previous paper (Templeton and Rothman, 1974) caution us about the dangers of trying to make too specific a definition of optimal. Instead we opt for a weaker definition of optimal which allows us to order fitness responses. We will say fitness response I is better than II if I increases the chances for existence at genetic equilibrium of its bearers or increases its frequency in the population if already present over that of II. Note that this definition of optimal already includes the impact of the genetic constraints and, because of the role played by higher moments, does not necessarily commit us to maximizing average fitness. Most importantly, we define 'optimality' in terms of the evolutionary fate of the population rather than the other way around in which the evolutionary fate is defined in terms of the optimality criterion.

In this section we will work with the general one-locus, two-allele model for a random mating diploid population with discrete generations:

	Genotype	**AA**	**Aa**
(3-1)	Fitness $= W_i$	$C_1 e^{[-\lambda\Sigma\delta_i(d_1)(n_i-d_1)]}$	$C_2 e^{[-\mu\Sigma\delta_i(d_2)(n_i-d_2)]}$
	Genotype	**aa**	
	Fitness	$C_3 e^{[-\nu\Sigma\delta_i(d_3)(n_i-d_3)]}$	

Where

$$\delta_i(d) = \begin{cases} 0 \text{ if } n_i \leqslant d \\ 1 \text{ if } n_i > d. \end{cases}$$

and i indexes all runs of 1's as before.

The allele frequency process is governed by:

$$q_{j+1} = [W_3(j)q_j^2 + W_2(j)q_j(1-q_j)]/[W_1(j)(1-q_j)^2 + W_2(j)2q_j(1-q_j) + W_3(j)q^2]. \tag{3-2}$$

As is easily seen, the situation discussed in the previous section is a special case of the above. In order to completely define (3-2), it is only necessary to specify the relative fitnesses. These are obtained by dividing the 'absolute' fitnesses given in (3-1) by W_2, the heterozygote fitness. Thus, the heterozygote always has a relative fitness of 1. With an environmental model as in (2-1) the expected logs of the relative fitnesses are:

(3-3)	Genotype	$E(\log W_{\text{rel}})$
	AA	$\log C_1 - \log C_2 - \lambda L p_1(1-\beta)^{d_1} + \mu L p_1(1-\beta)^{d_2}$

(3-3) **Aa** 0

aa $\log C_3 - \log C_2 - \nu L p_1 (1-\beta)^{d_3} + \mu L p_1 (1-\beta)^{d_2}$.

The variance in relative fitness is found from equation (2-4) by setting $w_n = w_n(\mathbf{AA})/w_n(\mathbf{Aa})$ for the relative fitness of the **AA** homozygote and $w_n = w_n(\mathbf{aa})/w_n(\mathbf{Aa})$ for the relative fitness of the **aa** homozygote. In terms of the three series given by (2-7) whose sums appear in (2-4), the appropriate sums needed to evaluate the variance of the log relative fitness of **AA** are:

$$S_1 = -\lambda(1-\beta)^{d_1}/\beta^2 + \mu(1-\beta)^{d_2}/\beta^2 = S_1(\mathbf{AA}) - S_1(\mathbf{Aa})$$

(3-4) $$S_2 = S_1(\mathbf{AA})(d_1 - 1 + 2/\beta) - S_2(\mathbf{Aa})(d_2 - 1 + 2/\beta) = S_2(\mathbf{AA}) - S_2(\mathbf{Aa})$$

$$S_3 = \lambda^2(1-\beta)^{d_1}(2/\beta - 1)/\beta^2 + \mu^2(1-\beta)^{d_2}(2/\beta - 1)/\beta^2 - 2\mu\lambda (1-\beta)^{\max(d_1, d_2)}(2/\beta - 1 + |d_1 - d_2|)/\beta^2 = S_3(\mathbf{AA}) + S_3(\mathbf{Aa}) - 2\mu\lambda(1-\beta)^{\max(d_1, d_2)}(2/\beta - 1 + |d_1 - d_2|)/\beta^2$$

where $S_i(\mathbf{AA})$ and $S_i(\mathbf{Aa})$ represent the sums given in (2-7) evaluated with the absolute fitnesses given in (3-1) for **AA** and **Aa** respectively. Similar formulae hold for the sums that appear in variance expressions of the log relative fitness of the **aa** genotype.

Using the results in Karlin and Lieberman (1974), we can conclude that fixation of the **A** allele will be stable if:

(3-5) $$\log C_1 - \log C_2 > L p_1 [\lambda(L\beta)^{d_1} - \mu(1-\beta)^{d_2}]$$

and fixation of the **a** allele will be stable if

(3-6) $$\log C_3 - \log C_2 > L p_1 [\nu(1-\beta)^{d_3} - \mu(1-\beta)^{d_2}].$$

If neither (3-6) holds, we have a polymorphic situation (although there may not be a stable polymorphic allele frequency – see Hartl and Cook, 1975). The only caution is that these arguments do not apply to the case when $C_1 = C_2 = C_3$, $\lambda = \mu = \nu$ and $d_1 = d_2 = d_3$. In this case, (3-5) and (3-6) are violated, but from (3-4) Var $(W_{\text{rel}}) = 0$ and these conclusions are valid only when Var $(W_{\text{rel}}) > 0$. Also, a stable polymorphism can exist even when (3-5) and (3-6) are satisfied providing the variance of the relative fitness is large. Unfortunately, the exact conditions necessary to produce this situation are unknown for the general case given by (3-1) (Karlin and Lieberman, 1974). However, the exact stability requirements have been worked out for the special case in which both homozygotes have equal relative fitnesses; i.e., when $\lambda = \nu$, $d_1 = d_3$ and $C_1 = C_3$. A possible example of this situation is also provided by maize. Two other alleles that have been isolated at the Adh_1 locus are $Adh_1{}^F$, an allele which codes for a relatively unstable but active polypeptide, and $Adh_1{}^{C(m)}$, which codes for a very stable but relatively inactive polypeptide (Schwartz and Laughner, 1969). However, the enzyme alcohol dehydrogenase is a 'dimer' which consists of an association of two polypeptide chains. In $Adh_1{}^F/Adh_1{}^{C(m)}$ heterozygotes about half of the dimers consists of an '*F*' polypeptide associated with a '*C*(*m*)' polypeptide. The resulting hybrid dimer is both stable and active, properties which neither homozygote has. Hence $Adh_1{}^F/Adh_1{}^{C(m)}$ individuals should be more buffered against runs of flooding conditions than either homozygote, although this prediction has not yet been tested.

When both homozygotes have equal fitness parameters, fixation for either the **A** or **a** allele is stable whenever:

$$\text{(3-7)} \qquad \log C_1 - \log C_2 > Lp_1\,[\lambda(1-\beta)^{d_1} - \mu(1-\beta)^{d_2}]\,.$$

Letting W be the relative fitness of the homozygote, $q^* = 1/2$ is also a stable solution whenever

$$\text{(3-8)} \qquad E\left[\log \frac{2W}{1+W}\right] < 0.$$

By using a Taylor's series expansion on $\log(1+e^{\ln W})$, (3-8) can be approximated by:

$$\text{(3-9)} \qquad E(\log W) < \mathrm{Var}\,(\log W)/4.$$

Thus, $E(\log W) < 0$ is a sufficient condition to insure both stability at $q^*=1/2$ and instability around 0 or 1. We therefore have a stable polymorphism and no other stable solution whenever (3-7) is violated.

We now draw our attention to the types of fitness responses which allow the heterozygotes to be present at equilibrium, i.e., 'optimal' responses on the part of the heterozygotes. We first consider the special case when $d_1 = d_2$. Under this assumption, a polymorphism is the only stable solution whenever:

$$\text{(3-10)} \qquad (\lambda-\mu) > \Delta/\mathrm{B}$$

where $\Delta = \log C_1 - \log C_2$ and $B = Lp_1(1-\beta)^d$ = the expected number of 'bad' time units encountered per generation. The parameter Δ measures the fitness difference in the inherent fitness component. Increasing Δ represents a strategy on the part of the heterozygote which says to put energy into those aspects of survival and reproduction which are totally unrelated to the environmental fluctuation. Such a strategy will not result in evolution of homeostasis to the environmental heterogeneity under consideration since it, in effect, ignores it. The parameter $(\lambda-\mu)$ however does measure the relative homeostatic capabilities of the two different genotypes to withstand runs of 1's. As this quantity increases, the heterozygote becomes more and more

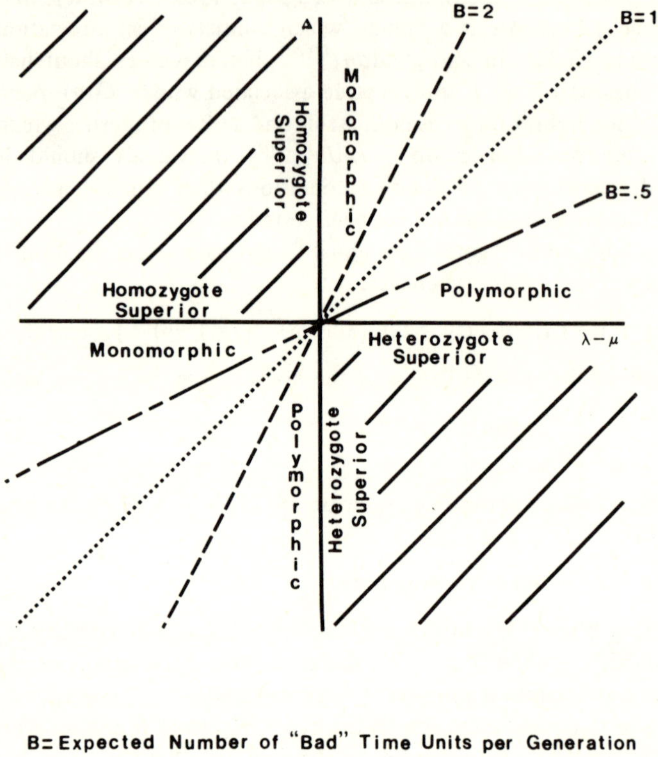

Fig. 2. The conditions under which evolution results in a polymorphic versus monomorphic population as a function of B, the expected number of 'bad' time units encountered per generation. Both heterozygotes and homozygotes have equal d parameters, but may differ in Δ, a measure of their inherent fitness that is unrelated to the environmental heterogeneity, and in their rate of fitness decline once in a run of 1's greater than length d. This rate of fitness decline may be interpreted as a measure of homeostasis. The difference in these rates between the homozygote (λ) and the heterozygote (μ) are plotted along the abscissa. All possible parameter values of Δ and $\lambda-\mu$ are assumed to lie in the upper right and lower left quadrants since by assumption no single genotype can have simultaneous superiority in both inherent fitness and homeostasis. The broken and dotted lines correspond to the boundary separating parameter values that yield polymorphic populations (below the lines) versus monomorphic populations (above the lines) as a function of B.

homeostatic relative to the homozygote. The importance of homeostasis to the heterozygote as a function of the environment is graphically portrayed in Figure 2. In this figure the lines $\lambda - \mu = \Delta/B$ are plotted for various B's. In the area on and below these lines a stable polymorphism is the only equilibrium solution and so heterozygote survival is insured. Above the lines monomorphic populations fixed for one of the alleles are stable. These areas are labeled monomorphic. Part of the 'monomorphic' area could include stable polymorphisms if (3-9) is satisfied, but even in these cases heterozygote survival is not insured because there are also two monomorphic stable solutions. Moreover, if we consider that one of the alleles had to have originated from the other by mutation, its initial frequency would be very small and it would thus be in the neighborhood of a stable monomorphic solution. For these reasons, we only label those areas as 'polymorphic' which are unambiguously polymorphic. Furthermore, we exclude the upper left hand and lower right hand quadrants. This derives from our no free lunch assumption; that is, no single organism can have both maximal inherent fitness and maximal homeostasis. Thus, because the organism has only so much energy to devote to fitness strategies, Δ and $\lambda - \mu$ must have the same sign. In examining Figure 2 we first confine our attention to the upper right hand quadrant. In this quadrant, the heterozygote has superior homeostatic capabilities, but inferior inherent fitness (those aspects of fitness not related to the environmental heterogeneity under consideration). As B increases the conditions which allow heterozygotes to survive greatly broaden. Similarly, in the lower left-hand quadrant heterozygotes which opt for a strategy of higher inherent fitness but lower homeostasis are 'squeezed' out as B increases. Consequently, we have the not too surprising result that devoting energy to homeostatic strategies becomes more and more optimal as the environment against which it buffers is encountered more and more often. B in turn increases as the average frequency of environment 1 increases, the runniness of the environment increases, d decreases or lifespan increases. Thus, we would expect organisms with longer lifespans in general to put more energy into homeostatic mehcanisms than those with short lifespans. In this regard it is interesting to note that mammals are generally the longest lived land animals and also are the ones with the most elaborate homeostatic mechanisms of both a physiological and behavioral nature.

Now consider the special case where $\lambda = \mu$. The conditions that insure a polymorphic equilibrium population are:

$$\text{(3-11)} \quad \frac{\Delta}{\lambda B} < 1 - (1 - \beta)^{d_2 - d_1}$$

where B= the expected number of bad time units per generation encountered by the homozygote. This inequality can also be graphed as was (3-10), and the result is shown in Figure 3. Once again, populations on or below the lines are polymorphic, but populations above the line do not necessarily have heterozygotes present at equilibrium. The lines are drawn for several values of $1-\beta$. As can be seen, if the environments are not too runny ($1-\beta$ small), the conditions for maintaining a polymorphism are greatly broadened by a slight heterozygote superiority in the d's, but the fitness return for increasing d

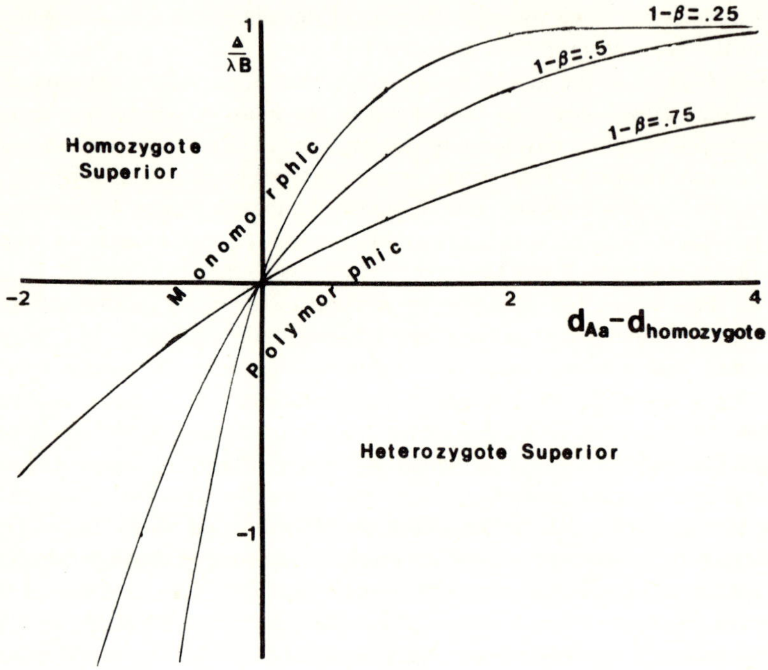

Fig. 3. The conditions under which evolution results in a polymorphic versus monomorphic population as a function of the runniness $(1-\beta)$ of the environment. The heterozygotes are sentivie to runs of 1's only of length d_{Aa} or greater, while homozygotes are sensitive to runs of length $d_{homozygote}$ or greater. Once into a run of 1's of sufficient length, both genotypes have identical fitness loss rates (λ), but may differ in their inherent fitnesses (Δ). Hence, all homeostatic differences between the genotypes in this case are measured by $d_{Aa}-d_{homozygote}$. The curved lines correspond to the boundary separating parameter values that yield polymorphic populations (below the curves) versus monomorphic populations (above the curves) as a function of $1-\beta$.

rapidly levels off. On the other hand, if the environment is very runny, a small increase in d yields only a small broadening of polymorphic conditions, but the advantage in increasing d now levels off only very slowly. Consequently, an optimal fitness strategy in an environment that is not very runny is to increase d slightly. The quantity 'd' measures how long the organism can withstand state 1 without any ill effects whatsoever. Thus, if the environment is not too runny we would expect organisms to evolve homeostatic solutions which effectively 'hide' them from potentially adverse environments for a short period of time. Of course, this homeostatic strategy becomes less effective as the environment becomes progressively more runny, as Figure 3 shows. Under these environmental conditions it might perhaps be better to achieve homeostatis through a small λ value since this would represent a long term solution. For example, suppose an organism is exposed to dry weather conditions. One way of protecting itself from dessication is to hide in a burrow or some shelter until the danger of dessication passes. Such homeostatic behavior is only effective if the dry spells do not last too long. If such dry periods last a long time, the organism will be forced to leave its burrow to engage in other life sustaining activites, and hence has the need to develop some long-time homeostatic solution, such as increased kidney efficiency, reduced transpiration losses from membranes, etc.

To investigate more carefully this prediction that 'd-type' (short-term) homeostatic mechanisms are more favored in unrunny environments, we must allow changes both in λ and d, and not fix one or the other as we have done in the above paragraphs. With our model, expected log fitness is expressed as

$$\text{(3-12)} \quad \log C - Lp_1(1-\beta)^d.$$

In general, factors which increase expected log fitness also increase chances for survival at equilibrium, both for the models considered in this section and in the previous one. However, because higher moments do enter into the determination of the quilibrium population, this is not universally true and exceptions can be constructed. We will ignore this complication and now assume that maximizing (3-12) is 'optimal' for the organism. To see how an organism can achieve optimal fitness by different homeostatic strategies, (3-12) is differentiated with respect to $-\lambda$ and d:

$$\text{(3-13)} \quad \frac{\partial E(\log W)}{\partial(-\lambda)} = Lp_1(1-\beta)^d$$

$$\text{(3-14)} \quad \frac{\partial E(\log W)}{\partial d} = -\lambda Lp_1(1-\beta)^d/\log(1-\beta).$$

The above equations show how a change in either homeostatic parameter causes a change in fitness. These derivatives already confirm the obvious; that is, decreasing λ increases fitness and increasing d increases fitness. To get a measure of the effectiveness in increasing fitness by decreasing λ versus increasing d, the ratio of (3-13) to (3-14) is taken:

$$(3\text{-}15) \quad R = \log(1-\beta)/(-\lambda).$$

When this ratio is greater than one, changing d by a given amount causes a greater increase in fitness than a corresponding change in λ, and the opposite is true when the ratio is less than done. In other words, when

$$(3\text{-}16) \quad -\log(1-\beta) < \lambda$$

it is more optimal to put energy into decreasing λ rather than increasing d. As the environment becomes more runny, $1-\beta$ approaches one and its log approaches 0. Consequently, for runny environments we would expect (3-16) to be satisfied over a large range of λ values. However, when the environment is not runny, the right hand side of (3-16) becomes larger and thus it is more likely for an increase in d to be an optimal homeostatic strategy. This confirms our prediction that short-term homeostatic mechanisms tend to be favored in unrunny environments while long-term solutions are favored in runny environments.

4. Spatial Heterogeneity, Habitat Selection and Environmental Manipulation

We continue our format of fixing three of the major model components and allowing the fourth to vary by now allowing the environmental model described by (2-1) to vary. The matrix of form (2-1) is our essential description of temporal heterogeneity in the environment, and by allowing (2-1) to vary we can now overlay spatial heterogeneity upon the temporal fluctuations governed by a particular environmental matrix. For example, we can imagine that the population of interest occupies a given area in space. At any particular location in this space, the temporal heterogeneity is described by a matrix of the form (2-1) but different localities in this area may have different values for α and β. Thus, there may be spatial heterogenity within the area occupied by this population with respect to average environmental state frequencies, runniness or both. This spatial heterogeneity can be described by a probability distribution over α and β. We further assume the spatial heterogeneity is continuous over the area under consideration, and hence we use continuous densities to describe the probability of picking out a particular α or β. Since α and β must be between 0 and 1, a convenient choice for the marginal distributions of α and β are beta distributions:

$$(4\text{-}1)\qquad \begin{aligned} f(\alpha) &= \frac{\Gamma(M)}{\Gamma(a_1)\,\Gamma(a_2)} \cdot \alpha^{a_1-1}(1-\alpha)^{a_2-1} \quad 0 \le \alpha \le 1 \\ f(\beta) &= \frac{\Gamma(N)}{\Gamma(b_1)\,\Gamma(b_2)} \cdot \beta^{b_1-1}(1-\beta)^{b_2-1} \quad 0 \le \beta \le 1 \end{aligned}$$

where $M = a_1 + a_2$ and $N = b_1 + b_2$. With these marginal distributions

$$(4\text{-}2)\qquad \begin{aligned} E(\alpha) &= a_1/M \\ E(\beta) &= b_1/N \\ \mathrm{Var}\,(\alpha) &= a_1 a_2/[M^2(M+1)] \\ \mathrm{Var}\,(\beta) &= b_1 b_2/[N^2(N+1)]\,. \end{aligned}$$

However, α and β are not in general independent, so let ρ be the correlation between them. Because of possible correlations, the joint distribution of α and β, $f(\alpha, \beta)$, is not necessarily the product of the marginals.

Assume a single random-mating, diploid plant or animal population inhabits the area described by $f(\alpha, \beta)$. We further assume that the offspring are dispersed at random across this space, and hence their distribution is also given by $f(\alpha, \beta)$. After this initial dispersal phase, the offspring settle in a particular locality and grow and reproduce under the environmental conditions present at this locality [some realization from $f(\alpha, \beta)$]. (Note, in Levins' terminology, we have overlaid 'coarse-grained' heterogeneity on top of 'fine-grained' heterogeneity; we ignore these considerations of grain for now, but will return to them in the discussion.) If the fitness has the same form as given by (2-2) and (2-6), the expected log fitness and its variance for an individual with genotype j given a particular set of (α, β) is

$$(4\text{-}3)\qquad \begin{aligned} E(\log\, W_j|\alpha,\beta) &= \log C_j - \lambda_j L\alpha(1-\beta)^{d_j}/(\alpha+\beta) \\ \mathrm{Var}\,(\log\, W_j|\alpha,\beta) &= \lambda_j^2\, L\alpha(1-\beta)^{d_j}\{2/\beta - 1 + \alpha(1-\beta)^{d_j}[2\alpha/(\beta\alpha+\beta^2) \\ &\quad -4/\beta - 2d_j + 1]/(\alpha+\beta)\}/(\alpha+\beta). \end{aligned}$$

Since we assume the population represents a single random-mating population over this entire area, the fitness process which governs the allele frequency changes and hence the evolutionary fate of the population is the fitness for *all* individuals of genotype j irrespective of the particular locality and one individual occupies. The mean of the log fitness for genotype j irrespective of locality is:

$$(4\text{-}4)\qquad E(\log W_j) = \int_0^1 \int_0^1 E(\log W_j/\alpha, \beta) f(\alpha, \beta)\, d\alpha\, d\beta$$

and the variance is:

$$(4\text{-}5)\qquad \mathrm{Var}(\log W_j) = \int_0^1 \int_0^1 \mathrm{Var}(\log W_j | \alpha, \beta) f(\alpha,\beta)\, d\alpha\, d\beta + \int_0^1 \int_0^1 [E(\log W_j | \alpha, \beta)]^2 f(\alpha, \beta)\, d\alpha\, d\beta - [E(\log W_j)]^2.$$

Equations (4-4) and (4-5) reflect the fact that there is a mixture of fitness processes for genotype j as determined by the nature of the spatial heterogeneity.

In order to examine the impact of spatial heterogeneity in runniness we now consider a special case of (4-1) in which the average state frequencies (p_1 and p_0) are held constant but the runniness (as measured by $1-\beta$) of the environment is allowed to vary. Such a situation is described by the probability distribution:

$$(4\text{-}6)\qquad f(\beta) = \frac{\Gamma(N)}{\Gamma(b_1)\,\Gamma(b_2)} \cdot \beta^{b_1-1}(1-\beta)^{b_2-1}.$$

Since $\alpha = \frac{p_0}{1-p_0}\beta$ in this case, $\rho = 1$ and it is not necessary to treat both α and β as random variables. Substituting (4-6) into (4-4) and (4-5), one obtains:

$$E(\log W_j) = \log C_j - \lambda_j L p_1 \frac{\Gamma(N)\Gamma(b_2+d_j)}{\Gamma(N+d_j)\Gamma(b_2)}$$

$$(4\text{-}7)\qquad \mathrm{Var}(\log W_j) = \lambda_j^2 L p_1 \frac{\Gamma(N)\Gamma(b_2+d_j)}{\Gamma(N+d_j)\,\Gamma(b_2)} \left\{ \frac{2(N+d_j-1)}{b_2-1} - 1 + p_1 \frac{\Gamma(b_2+2d_j)\Gamma(N+d_j)}{\Gamma(b_2+d_j)\Gamma(N+2d_j)} \times \left[\frac{(2p_1-4)(N+2d_j-1)}{b_1-1}\right]\right\} - 2d_j + 1 + \lambda^2 L^2 p_1{}^2 \left[\frac{\Gamma(N)\Gamma(b_2+2d_j)}{\Gamma(N+2d_j)\Gamma(b_2)} - \left(\frac{\Gamma(N)\Gamma(b_2+d_j)}{\Gamma(N+d_j)\Gamma(b_2)}\right)^2\right].$$

We now impose some genetic constraints upon this model. Suppose we have the one locus 'recessive' case given in Section 2. We can now consider the evolutionary situation that arises when dealing with an environment with no spatial heterogeneity versus the situation with spatial heterogeneity but with identical average runniness ($1-\beta$). Some specific examples of these contrasts are given in Table 4-1. Some general patterns emerge upon inspection

TABLE 4-1

The effect of spatial heterogeneity upon the mean and variance of log fitness. In all cases $\log C = 0.1$, $\lambda = 0.1$, $d = 5$, $L = 100$ and $p_1 = 0.25$. The genetic constraints are like those given in Section 2 in which the **A** allele is dominant over the **a** allele.

Spatial Heterogeneity	$E(1-\beta)$	b_1	b_2	$E(\log W)$	Var $(\log W)$	Type of equilibrium population
No	0.75	–	–	–0.493	0.331	Fixed for **A**
Yes	0.75	2	6	–0.695	1.759	Polymorphic
No	0.50	–	–	0.022	0.019	Fixed for **a**
Yes	0.50	4	4	–0.077	0.156	Polymorphic
No	0.625	–	–	–0.138	0.093	Fixed for **A**
Yes	0.625	3	5	–0.298	0.390	Fixed for **A**

of this table. First, spatial heterogeneity always causes a decrease in expected log fitness over the case with the same average environment but no spatial heterogeneity. Second, the variance in log fitness is always increased when spatial heterogeneity is introduced. The increase in variance is always much greater proportionally than the decrease in the mean log fitness. As discussed in Section 2, the variance can be an important determinant of the equilibrium population, and this latter observation implies that the variance plays an even more critical role when spatial heterogeneity is present. In particular, we saw in Section 2 that a large variance makes it more likely for the equilibrium to be polymorphic. That this is true is easily seen from Table 4-1.

These same conclusions are true under other types of genetic constraints. Consider the genetic model given by (3-3) with equal homozygote fitnesses. Some special cases of this are given in Table 4-2. Once again, spatial heterogeneity decreases the relative fitness of the genotype with the lesser homeostatic ability. Consequently, spatial heterogeneity increases selection favoring homeostasis even though with this model any particular individual does not experience this spatial heterogeneity within its own lifetime. Also, spatial heterogeneity greatly increases the variance of the log relative fitness. When the heterozygote is the more homeostatic (which is usually found to be empirically true, e.g., see Lerner, 1954), the impact of spatial heterogeneity is to increase the fitness of the heterozygote relative to the homozygote and

TABLE 4.2.

The effect of spatial heterogeneity upon the mean and variance of log fitness. In all cases $d=5$, $L=100$, and $p_1=0.25$. The genetic constraints are like those given in Section 3 in which the heterozygote has one fitness response and both homozygotes share another.

Spatial Heterogeneity	Δ	$\mu-\lambda$	$E(1-\beta)$	b_1	b_2	$E(\log W_{rel})$	Var $(\log W_{rel})$	Type of equilibrium population
No	0.25	0.1	0.75	–	–	–0.343	0.331	Polymorphic
Yes	0.25	0.1	0.75	2	6	–0.545	1.759	Polymorphic
No	0.25	0.1	0.625	–	–	0.012	0.093	Potentially Polymorphic
Yes	0.25	0.1	0.625	3	5	–0.148	0.390	Polymorphic
No	0.25	0.1	0.50	–	–	0.172	0.019	Mono-morphic
Yes	0.25	0.1	0.50	3	3	0.042	0.231	Potentially Polymorphic
No	0.20	0.1	0.50	–	–	0.122	0.019	Mono-morphic
Yes	0.20	0.1	0.50	3	3	–0.008	0.231	Polymorphic
No	–0.20	–0.1	0.50	–	–	–0.122	0.019	Polymorphic
Yes	–0.20	–0.1	0.50	4	4	–0.023	0.156	Polymorphic
Yes	–0.20	–0.1	0.50	3	3	0.008	0.231	Potentially Polymorphic
No	–0.36	–0.1	0.75	–	–	0.233	0.331	Mono-morphic
Yes	–0.36	–0.1	0.75	2	6	0.435	1.759	Potentially Polymorphic
No	–0.25	–0.1	0.625	–	–	–0.012	0.093	Polymorphic
Yes	–0.25	–0.1	0.625	3	5	0.148	0.390	Mono-morphic

to increase the variance. Both of these phenomenon tend to increase the chances for a polymorphic population (see conditions 3-7 and 3-9). In the case when the homozygote is more homeostatic, these two phenomenon are in opposition in the manner in which they affect the quilibrium population. Consequently, sometimes spatial heterogeneity favors polymorphism, sometimes it does not when homozygotes have superior homeostatic capabilities. The fact that spatial heterogeneity can actually make a monomorphic situation turn into a potentially polymorphic one (Table 4-2) is particularly interesting when one considers that the mean log relative fitness of the heterozygote has actually decreased, yet now the heterozygote can exist where it could not before. Consequently, we conclude that spatial heterogeneity generally favors maintenance of polymorphism through increasing the variance of the fitness process.

Up to now we have assumed the offspring are distributed at random over the area each generation. However, suppose certain genotypes have the ability to 'select' certain habitats, i.e., they are found in certain localities with greater than expected frequency and other localities with lesser than expected frequency. To investigate the possible implications of habitat selection (non-random dispersal over the available area), we now assume that all genotypes have identical fitness responses to a given set of temporal states. In other words, $\lambda_j=\lambda$ for all genotypes j, $d_j=d$ for all j and $\log C_j = \log C$ for all j (for convenience, see set $\log C=0$ in this section). All these genotypes are assumed to belong to a single random mating population with discrete generations inhabiting a spatially heterogeneous area as described by (4-6). Some of the genotypes disperse at random over this area before settling down and hence their distribution is also given by (4-6). However, other genotypes do not disperse according to (4-6), but with the distribution:

$$g_j(\beta) = \frac{\Gamma(V_j)}{\Gamma(\nu_{j1})\Gamma(\nu_{j2})} \cdot \beta^{\nu_{j1}-1}(1-\beta)^{\nu_{j2}-1} \tag{4-8}$$

where the subscript j indicates this distribution is a function of the genotype j. Consequently, although all organisms have the same fitness response parameters, they do have the option of changing their fitness by choosing their environments rather than randomly dispersing. We will always assume this choice is made so as to decrease $1-\beta$ and hence to increase expected log fitness (the possible dangers of this as an optimality criterion have already been discussed in Section 3). However, the no free lunch assumption says that exercising this option of habitat selection involves some cost to the organism. For example, the habitat selector in order to make a selection must first sample some part of the area and then choose among the sampled areas. It takes energy to make a sample, and furthermore the organism is usually more exposed to predation during this state. Also, in order to make a choice, the organism must have the neurological mechanisms necessary to store and analyze information. Maintaining these mechanisms would generally subtract from other fitness components. Consequently, we assume that the fitness of genotype j is of the form:

$$W_j = H_j e^{-\lambda B[d,\, g_j(\beta)]}$$

where H_j represents the 'cost' to the genotype of changing its distribution from $f(\beta)$ (4-6) to $g_j(\beta)$, and $B[d, g_j(\beta)]$ is the number of 'bad' time units which is a function of both d and $g_j(\beta)$. In general, we will assume the cost increases (H_j decreases) as $B[d, g_j(\beta)]$ decreases because this would biologically mean that the organism is putting more time and energy into habitat selection. In this model the only fitness differences that arise between genotypes are those due to habitat selection.

We first examine what happens to fitness ignoring the cost of habitat selection. If there was no cost to habitat selection the expected log fitness of genotype j would be: (letting log C=0)

$$(4\text{-}9) \qquad -\lambda L p_1 \, \Gamma(V_j)\Gamma(v_{2j}+d)/[\Gamma(V_j+d)\Gamma(V_{j2})] \, .$$

Without habitat selection, the expected log fitness would be

$$(4\text{-}10) \qquad -\lambda L p_1 \, \Gamma(N)\Gamma(b_2+d)/[\Gamma(N+d)\Gamma(b_2)] \, .$$

Consequently, if there was no cost, genotypes with habitat selection would have higher expected log fitnesses whenever:

$$(4\text{-}11) \qquad \frac{\Gamma(V_j)\Gamma(V_{2j}+d)}{\Gamma(V_j+d)\Gamma(v_{2j})} < \frac{\Gamma(N)\Gamma(b_2+d)}{\Gamma(N+d)\Gamma(b_2)} \, .$$

Thus, both the mean and variance of the habitat selectors' distribution can influence its fitness. In the special case in which the constraint $V_j=N$ is imposed, (4-11) becomes:

$$(4\text{-}12) \qquad v_{2j} < b_2$$

that is, whenever the average runniness of the environment is decreased by the act of habitat selection. Note also that when (4-12) holds, Var $[\beta \,|\, g_j(\beta)]$ < Var $[\beta \,|\, f(\beta)]$, and Var (log W_j) < Var (log W). The variance of the log fitness is further reduced when $V_j > N$ is true as well as (4-12). As we saw earlier, spatial heterogeneity tends to increase the importance of the variance of log fitness in determining the equilibrium population. However, as we see here habitat selection has the potential for reducing the impact of the spatial variability.

Suppose the constraint $V_j=N$ holds for a certain population, so that all habitat selection is determined by changes in v_{2j}; i.e., by changing the mean runniness of the environment. Let the cost of changing the mean runniness from $1-\bar{\beta}_r$ under random dispersal to $1-\bar{\beta}_j$ under habitat selection be of the form:

$$\text{(4-13)}\quad \log H_j = -c[1-\bar{\beta}_r-(1-\bar{\beta}_j)] = -c(\bar{\beta}_j-\bar{\beta}_r) = \frac{-c(\nu_{1j}-b_1)}{N}$$

where c=cost rate. Consequently, the more the genotype shifts the mean runniness, the greater the fitness cost of habitat selection. The expected log fitness is now:

$$\text{(4-14)}\quad -c(\bar{\beta}_j-\bar{\beta}_r)-\lambda Lp_1\Gamma(N)\Gamma(\nu_{2j}+d)/[\Gamma(N+d)\Gamma(\nu_{2j})].$$

Comparing this to (4-10), it can be seen that a genotype engaging in habitat selection will have a higher expected log fitness than a random dispersal genotype whenever:

$$\text{(4-15)}\quad c < \frac{-\lambda Lp_1\Gamma(N)[\Gamma(\nu_{2j}+d)/\Gamma(\nu_{2j})-\Gamma(b_2+d)/\Gamma(b_2+d)]/\Gamma(N+d)}{\bar{\beta}_j-\bar{\beta}_r}$$

This relationship is shown graphically in Figure 4. As can be seen, as c increases the range of conditions favoring habitat selection becomes increasingly narrow. A specific example of this is also shown on Figure 4. In this example, λ=0.1, L=100, d=5, p_1=0.25, N=8 and b_2=6 (giving an average runniness of 0.75). The difference between (4-9) and (4-10) (the expected log fitness if no costs were incurred during habitat selection is plotted against $\bar{\beta}_j-\bar{\beta}_r$, a measure of the effectiveness of habitat selection). However, when costs are considered only that portion of the resulting curve above the cost line represents a situation where habitat selection is favored. As is easily seen, the portion of this curve above these cost lines becomes smaller and smaller as c increases. Also shown on this graph are the variances in log fitness at various points along the curve. This demonstrates that these variances can decrease dramatically (in the example shown, 2 orders of magnitude) as habitat selection becomes more effective.

The portion of the curve above the cost lines shows when a genotype engaging in habitat selection is favored over random dispersing genotypes. However, other habitat selector genotypes may exist, and it then becomes important to see which has the highest expected log fitness. One can find the 'optimal' compromise between costs and returns by differentiating (4-14) with respect to ν_{j2} and setting it equal to zero. Doing so, one finds that the ν_{j2} that maximizes expected log fitness satisfied the relationship:

$$\Psi(\nu_{j2}+d)-\Psi(\nu_{j2}) = c/N \div \lambda Lp_1\Gamma(N)\Gamma(\nu_{j2}+d)/[\Gamma(N+d)\Gamma(\nu_{j2})] = c/N \div [-E(\log W|\text{no cost})]$$

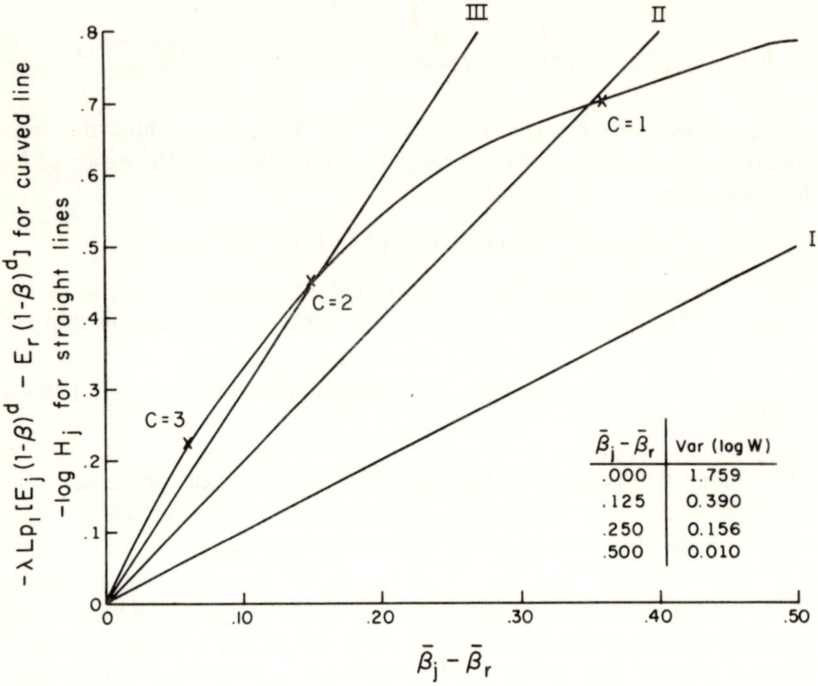

Fig. 4. The conditions under which habitat selection is favored. The effectiveness of habitat selection in reducing environmental runniness is plotted along the abscissa ($\beta_j - \beta_r$). The ordinate corresponds to both the costs accrued due to habitat selection (the straight lines) and to the increase in the fitness component sensitive to environmental runs that occurs as the effectiveness of habitat selection increases (curved line). The curved line shown corresponds to the following parameter values: $\lambda = 0.1$, $L = 100$, $d = 5$, $p_1 = 0.25$, $N = 8$ and $b_2 = 6$. Three different cost lines are plotted corresponding to cost rates of $c = 1$ (line I), $c = 2$ (II) and $c = 3$ (III). Only that portion of the curved line above the cost line for a specific cost rate corresponds to a situation in which the habitat selector has a higher fitness than the random disperser for a given cost rate. As can be seen, this portion decreases with increasing cost rate. Also, for each given cost rate there exists a point that maximizes expected log fitness. These points are indicated by an 'x' on the curved line with the appropriate cost rate printed by it. As the cost rate increases, these points shift down the curved line. Finally, the effect of habitat selection on the variance of log fitness is given by the table in the lower right hand corner. This shows that habitat selection can greatly decrease the variance.

where Ψ is the digamma function (Abramowitz and Stegun, 1965). The points on the curve drawn in Figure 4 that satisfy this for various c's are shown on Figure 4 by an 'x' with the value of c printed by it. As c decreases, it becomes more and more optimal to increase ν_{j2}.

Another interpretation of the above analysis is possible. Suppose the organism does not actually select habitats, but modifies the environment it inhabits so as to decrease the chance of having a run of 1's. The above model can be applied to this situation of environmental manipulation simply by regarding (4-8) as the distribution after manipulation. All other formulaes would be identical and the previous analysis could be used to study the evolutionary impact of this problem simply by substituting the words 'environmental manipulation' for 'habitat selection' in the preceding paragraphs.

Besides reducing the importance of the variance in log fitness, habitat selection and environmental manipulation can also influence the evolution of homeostasis. To see this, we now differentiate (4-14) with respect to $-\lambda$ and d to obtain:

$$\partial \log W/\partial(-\lambda) = Lp_1\Gamma(N)\Gamma(\nu_{2j}+d)/[\Gamma(N+d)\Gamma(\nu_{2j})]$$

$$\partial \log W/\partial d = \lambda Lp_1\Gamma(N)\Gamma(\nu_{2j}+d)\,[\Psi(N+d)-\Psi(\nu_{2j}+d)]/[\Gamma(N+d)\Gamma(\nu_{2j})]\,.$$

As the effectiveness of habitat selection or environmental manipulation increases (ν_{2j} increases), both of these derivatives decrease. This indicates that the fitness advantage of decreasing λ or increasing d (both of which increase 'homeostasis') decreases with increasing ν_{2j}. This is not surprising since habitat selection or environmental manipulation can also be regarded as homeostatic mechanisms.

In light of these conclusions, it is interesting to review a study done by Selander and Kaufman (1973). They noted that levels of genic heterozygosity, as measured by surveys of allelic variation of enzyme-coding loci, are much lower in populations of large, mobile animals (most vertebrates) than in those of small, relatively immobile animals (most invertebrates). They explained this phenomenon by arguing that most environmental heterogeneity is experienced by small animals as 'coarse-grained' and by large animals as 'fine-grained'. They then invoked Levins' (1968) prediction that polymorphism can exist only in coarse-grained environments. There are several difficulties with this argument. First, there is much ambiguity in the use of the word 'grain'. This is a point we will deal with in the discussion, so we now defer further comment on this until then. Second, Levins' statement about 'polymorphisms' applies *only* to the fitness set analysis with no genetic constraints. His conclusion has *nothing* to do with genic polymorphism. This point was made by Levins himself (1968) since he showed that when genetic constraints are considered, fine-grained environments can indeed yield genic polymorphism. Finally, as Templeton and Rothman (1974) point out, the conclusion that fine-grained environments yield only monomorphic 'optimal' populations in the absence of genetic constraints is itself suspect. In particular, an infinite number of optimality criteria can be constructed which choose a population from the admissible set, all of which can yield polymorphic optimum in fine-grained environments. Hence, Levins' conclusion is extremely sensitive to the precise form of the adaptive function and has no general validity even in terms of fitness set analysis. Consequently, fitness set analysis does not provide an explanation for the data surveyed by Selander and Kaufman. However, the model considered in this section does.

Small relatively immobile invertebrates have only limited opportunities for habitat selection. Furthermore, many invertebrates lack the degree of neurological complexity needed for effective habitat selection. On the other hand, most vertebrates have both the mobility and neurological apparatus necessary for effective habitat selection. Finally, as pointed out previously, there is much empirical evidence that heterozygotes generally have better homeostatic capabilities than homozygotes. Coupling these observations with the conclusions made in this section about the relation of habitat selection to the evolution of homeostasis, we would expect selection to result in genic polymorphism more often in immobile invertebrates than in mobile vertebrates; exactly the pattern that is observed in nature. The assumption of heterozygote superiority in homeostatic capability is not crucial to this prediction, for as we have previously demonstrated spatial heterogeneity, particularly when coupled with ineffective habitat selection, can greatly inflate the variance in fitness and thus in general makes genic polymorphism more likely even in the absence of heterozygote superiority. Moreover, it is interesting to note that birds generally have the lowest levels of genic polymorphism among the vertebrates (Selander and Kaufman, 1973). Birds, being very mobile, can survey large areas at little cost for the purpose of habitat selection. In terms of our model, this reduces the variance in fitness and thus reduces the chances for a genic polymorphism to be stable. Our model therefore explains the relatively low levels of genic polymorphism in birds, as well as the more general contrast between vertebrates and invertebrates.

5. Sex versus Parthenogenesis in Fine-Grained Environments

Up to now we have allowed all the major components of our fine-grained models to vary except for the genetic constraints. However, such factors as 'dominance' and mating structure are themselves subject to evolutionary change. In this section we allow variation in an even more fundamental genetic constraint – the mode of reproduction. In particular, we consider a diploid population which can either reproduce sexually with random mating or parthenogentically (eggs develop without male fertilization) by either an apomictic (loss of meiosis) or automictic (retention of meiosis) mechanism.

The evolution and retention of sex in many plant and animal species has long been a subject of debate (for review, see Williams, 1975, or Felsenstein, 1974). There are many negative fitness aspects to sexual reproduction. First, there are problems simply associated with the fact that a pair of organisms are required to reproduce rather than a single one. Time and energy must be spent in locating and attracting individuals of the opposite sex. More time and energy are required for the actual courtship, territorial defense and other mechanisms that insure the union of gametes. In addition, these activities often require considerable time in open, unsheltered areas where the risk of predation is great. Finally, a sexually reprodueing individual passes on only one-half of its genetic complement to each offspring, while an asexual or apomictic individual passes on all of its genes to each offspring (automictic individuals can be intermediate between the sexual and apomictic case, but usually will pass on nearly all their genes). Consequently, sexual reproduction has an inherent fifty percent disadvantage in relation to asexual reproduction from the genetic point of view (Maynard Smith, 1971).

Because of the disadvantages of sex, many biologists have tried answering the question, 'Why reproduce sexually?' Almost all the explanations that have been advanced to answer this question involve at some point environmental heterogeneity (Maynard Smith, 1971 and 1974; Williams and Mitton, 1973; Felsenstein, 1974; Nyberg, 1975; Williams, 1975). Although these models vary a great deal in the specifics, the basic idea is that sexual reproduction is advantageous when the environment experienced in the present generation is a poor predictor of future environments. Under these conditions, those individuals that produce genetically diverse offspring can make a larger genetic contribution to future generations than those which produce only genetically identical copies of themselves.

All of these models used to explain the evolution and retention of sexual reproduction involve 'coarse-grained' environmental heterogeneity; that is, the environmental variation is experienced between generations and not within it. Furthermore, these inter-generational changes often have to be very drastic (Maynard Smith, 1971) and engender intense selection (Milton and

Williams, 1973). However, in the models developed in this paper, the relevant environmental variation is experienced *within* the generation. To be sure differences do occur between generations because each generation only 'samples' a finite sequence of the environmental process, but by assumption the environmental process is governed by identical probabilistic rules each generation and L is large. Hence, the inter-generational environmental heterogeneity is not drastic nor does it engender intense selection changes.

How does such fine-grained heterogeneity affect the evolution and retention of sex? To gain some insight into this problem we will reconsider the fitness model given by (3-1) when homozygote fitnesses are identical. When this model is applied to a random-mating sexual population, a stable polymorphism at $q^*=1/2$ can occur when (3-8)

$$E(\log W_{\text{rel}}) < \text{Var}\,(\log W_{\text{rel}})/4$$

where W_{rel}= the relative fitness of the homozygote to the heterozygote = W_1/W_2. The solution $q^*=1/2$ is the only stable solution whenever

$$(5\text{-}1) \qquad E(\log W_{\text{rel}}) < 0 \quad \text{or} \quad E(\log W_1) < E(\log W_2).$$

When (5-1) holds the heterozygote on the average has the higher fitness. Otherwise, $q^*=0$ and $q^*=1$ are also stable solutions.

Suppose such a monoecious sexual population exists and is polymorphic at $q^*=1/2$ [(3-8) is satisfied]. Suppose now that a mutation occurs in one of the heterozygotes causing it to reproduce parthenogentically by apomixis. This mutant heterozygote will only produce heterozygous offspring and likewise its offspring will only produce heterozygotes, etc., etc. We asume no changes in the heterozygotes' response to runs of 1's occurs (λ and d are unchanged), but because of the elimination of some of the secondary disadvantages to sex the log C portion of the log fitness may increase by an amount Δ. Thus, letting a 'p' subscript designate a parthenogenetic heterozygote and a 's' subscript a sexual heterozygote,

$$(5\text{-}2) \qquad E(\log W_{2p}) = E(\log W_{2s}) + \Delta.$$

We now investigate the conditions under which the parthenogenetic heterozygotes can displace the sexual population. We cannot use relative fitness for this purpose because the apomictic and sexual populations are genetically isolated from one another as soon as the parthenogenetic mutation occurs. This genetic isolation means the parthenogenetic individuals make no contribution to an equation of the form (3-2), so relative fitness does not provide a 'common' fitness currency by which comparisons can be made.

Population size does provide a common currency. Population size can be related to fitness by regarding the fitness of (3-1) as 'absolute' fitness that indicates the actual number of adult progeny produced by an individual. Since the parthenogenetic population consists solely of heterozygotes and starts by assumption from a single individual, the number of parthenogenetic individuals at generation n is

$$(5\text{-}3) \qquad N_p(n) = \prod_{i=1}^{n} W_{2p}(i) = e^{\left(\sum_{i=1}^{n} \log W_{2p}(i)\right)}$$

where $W_{2p}(i)$ is the realized fitness at generation i and $\log W_{2p}(i)$ can be interpreted as the realized growth rate. This is the basic model of exponential 'density-independent' growth for discrete generations. It should be pointed out that growth does not always have to follow (5-3) at all times for these 'density-independent' analyses to be valid. For example, if the growth rates are positive, this model says the population size will become infinite. However, events could occur from time to time which limit population growth. (5-3) is still a valid model as long as the events which limit growth are external to the population and do not interact with the fitness response parameters (Andrewartha and Birch, 1954).

Since the sexual population is at allele frequency equilibrium, its growth is governed by (Fisher, 1958)

$$(5\text{-}4) \qquad N_s(n) = N_s(0)e^{\left(\sum_{i=1}^{n} \log(W_1(i)/2 + W_{2s}(i)/2)\right)}$$

where $W_1(i)/2 + W_{2s}(i)/2$ represents the realized 'average fitness' at generation i. Under the assumption of density-independent limitations on population growth, that population with the larger average growth rate will ultimately displace the other.

The average growth rates are obtained from (5-3) and (5-4) as $E(\log W_{2p})$ for the apomictic population and $-\log 2 + E[\log(W_1 + W_{2s})]$ for the sexual population. We are interested in the conditions that cause

$$(5\text{-}5) \qquad E(\log W_{2p}) > -\log 2 + E[\log(W_1 + W_{2s})]$$

that is, when the parthenogenetic population will replace the sexual one. By setting $W_{\text{rel}} = W_1/W_{2s}$, $\log(W_1 + W_{2s}) = \log(W_{\text{rel}} + 1) + \log W_{2s}$, using (5-2) and expanding $\log(W_{\text{rel}} + 1)$ by a Taylor's series, (5-5) can be re-expressed as

$$(5\text{-}6)\qquad 2\Delta - E(\log W_{\mathrm{rel}}) > \mathrm{Var}\,(\log W_{\mathrm{rel}})/4.$$

Inequality (5-6) is more likely to hold as the advantage of switching to parthenogenetic reproduction (Δ) increases or as the fitness of the heterozygote increases. In the special case where $\Delta=0$, a necessary but not sufficient condition for (5-6) to hold is the existence of a mean heterozygote advantage $[E(\log W_{\mathrm{rel}}) < 0]$, a condition that results in polymorphism being the only stable solution for the sexual population. Moreover, the right hand side of (5-6) represents the between generation fitness variability created by the finite sampling of the within generation environmental states. As the between generation variation increases, sexual reproduction is more likely to win out over parthenogenetic reproduction, a conclusion consistent with previous studies on the evolution of sex. Parthenogenesis will only win out when there is sufficient heterosis and/or a sufficient loss of the secondary disadvantages to sexual reproduction to overcome the between generation variation.

We now extend these results to the automictic case. Automixis is more complicated than apomixis because meiosis is retained. Because of meiosis, a heterozygous individual can produce genetically diverse progeny under automixis. This property can be described by the parameter 'K' (Templeton and Rothman, 1973) where K is the probability of a heterozygote producing automictically a heterozygous zygote. K in turn is a function of the exact mode of automixis and linkage to the centromere (Templeton and Rothman, 1973) and can range from 0 to 1 inclusively. Under automixis homozygotes only produce more homozygotes. Thus, under automixis the heterozygotes contribute to both the heterozygote and homozygote classes of the next generation, but homozygotes contribute only to the homozygote class. Thus the automictic heterozygotes represent a class genetically isolated from all other genotypes, whether sexual or parthenogenetic. We may take advantage of this genetic isolation in studying the evolutionary fate of automictic populations. There are two possible cases to consider. First, previous theory has shown heterozygotes can only exist in an automictic population if mean heterozygote superiority exists $E(\log W_{\mathrm{rel}} < 0)$ (Asher, 1970; Templeton and Rothman, 1973; Templeton, 1974). If such is not the case, the automictic population eventually becomes totally homozygous. Since homozygotes produce parthenogenetically only homozygotes, this situation is analogous to the apomictic case where all parthenogenetic progeny are genetically identical to their parent. Consequently, the apomictic analysis can be applied to this case except we now express (5-6) in terms of the relative fitness $W_2/W_{1s} = W'_{\mathrm{rel}}$ and $\Delta' = \log C_{1p} - \log C_{1s}$. In the case where heterozygotes persist in the automictic population, the growth of heterozygotes is given by

$$(5\text{-}7) \qquad N_{pAa}(n) = \mathrm{e}^{\left(\sum_{i=1}^{n} \log KW_{2p}(i)\right)}.$$

By assumption the parthenogenetic heterozygotes have a higher expected fitness than the parthenogenetic homozygotes, so if the parthenogenetic heterozygotes go extinct in competition with the sexual population the entire parthenogenetic population also goes to extinction. The parthenogenetic heterozygotes will persist whenever

$$(5\text{-}8) \qquad E(\log KW_{2p}) > E(\log W_1/2 + W_{2s}/2).$$

(5-8) can be re-expressed as

$$(5\text{-}9) \qquad \Delta + \log K - \log(1/2) > E[\log(1 + W_{\text{rel}})].$$

If there is no special advantage to parthenogenetic reproduction ($\Delta=0$), we can see from (5-9) that a necessary but not sufficient condition for the persistence of the parthenogenetic population is that $K > 1/2$; that is, the parthenogenetic heterozygotes must be producing more heterozygous offspring than the sexual heterozygotes. Using a Taylor's series, (5-9) can be approximated by:

$$(5\text{-}10) \qquad 2\,\Delta - E(\log W_{\text{rel}}) > \text{Var}\,(\log W_{\text{rel}})/4 - 2\log K.$$

Note that (5-6) is a special case of (5-10) with $K=1$. When K is less than one it is generally more difficult for the parthenogenetic population to persist. Consequently, the benefits of asexuality are reduced when meiosis is retained.

Heterozygote superiority in homeostatic capabilities could play an important role in the evolution of parthenogenetic populations if the within lifetime environment heterogeneity is such that selection for superior homeostasis is more important than selection for fitness components unrelated to the fine-grained heterogeneity (Section 3). Under these conditions, the superior homeostatic capabilities of the heterozygotes would both increase $-E(\log W_{\text{rel}})$ and decrease $\text{Var}\,(\log W_{\text{rel}})/4$ making it more likely that (5-10) or (5-6) would hold. In this regard it is interesting to note that there is much empirical evidence that heterozygotes drawn from randomly mating sexual populations generally have superior homeostatic capabilities (Lerner, 1954; Thoday, 1955; Lewontin, 1958; Koehn and Mitton, 1972). There are also many instances in which parthenogenetic populations have evolved from and then displaced their bisexual ancestors. For example, in 1965 Vanzolini (1970) found a bisexual population of the lizard *Cnemidophorus leminscatus* in Obidos, Brazil. However, by 1968 it had been completely replaced by an automictic parthenogenetic population. Seiler (1961, 1967) and Lokki *et al.*, (1975), have studied in detail the evolution of parthenogenetic races of the European bagworm, *Solenobia triquestrella*. The parthenogenetic populations of this species reproduce automictically by 'central fusion' with crossover suppression, a combination which results in $K=1$ (Carson, 1967). Indeed, Lokki *et al.* (1975) have shown parthenogenetic animals are highly heterozygous. In this bagworm, it is the parthenogenetic species that is widespread throughout Europe, while the bisexual ancestor is extinct except for a few small and highly isolated populations scattered over the Alps. Homeostasis also seems to play an important role in the maintenance of the only *Drosophila* species that is known to be totally parthenogenetic, *D. mangabeirai*. Murdy and Carson (1959) have shown this is an automictic species with crossover suppression ($K=1$). Furthermore, Carson (1962) did extensive studies on the natural habitats of this species in Central America and concluded the success of this species was due to the superior homeostatic properties of individuals.

These observations coupled with the theory developed in this section suggest one should qualify the statement that environmental heterogeneity favors sexual reproduction. This statement seems to be true only when the environmental variability is sufficiently 'coarse-grained' (between generations), but when dealing with fine-grained heterogeneity asexual reproduction can often be favored because it can more efficiently maintain homeostatic genotypes.

6. Discussion

The title of this paper includes the phrase 'fine-grained' because we want to emphasize the fact we are dealing with environmental heterogeneity that is experienced within the lifetime of an individual. In the introduction, fine-grained environments were defined as environments experienced as a succession of possibly different states by an individual organism within its own lifetime, and, moreover, the environmental process is stationary over time (generations). The opposite extreme is a 'coarse-grained' environment in which any particular individual lives out the selectively relevant portion of its life in just one environmental state, but heterogeneity exists at the population level in either space and/or time between generations. These definitions are the ones originated by Levins (1968), and have appeared in numerous papers since then (e.g., Templeton and Rothman, 1974; Selander and Kaufman, 1973; Williamson and Nelson, 1972; Gillespie, 1974; Strobeck, 1975;

and Emlen, 1973). At first these definitions seem clear cut, but actually many ambiguities are contained within them, and the models developed in this paper can be used to demonstrate these ambiguities.

The first ambiguity resides in the expression of 'experiencing' an environment. To illustrate the importance of what we mean by 'experiencing', let us return to the model developed in Section 2 and in particular to the examples that were graphically portrayed in Figure 1. In our first example from that section we had two environmental states, '0' and '1), which were governed by the transition matrix:

$$\begin{pmatrix} 5/8 & 3/8 \\ 3/4 & 1/4 \end{pmatrix}$$

and the individuals sampled 100 time units from this process. The expected proportion of an individual's lifetime spent in state '0' is 2/3, while the variance is 0.00173 (using Good's results, 1973). The low variance means the organisms experience both 0's and 1's within their own lifetime close to the average state frequencies. It is very unlikely to experience all 100 units as 0's or 1's, or even come close to that situation. Consequently, individuals do indeed experience a succession of 0's and 1's within their lifetime, and the model seems to be unambiguously fine-grained. However, in reformulating this model into Levins' fitness set format, we had to redefine our 'environmental states'. We went from 0 and 1 to 'good' and 'bad' where a 'bad' time unit is defined to be a state 1 that was preceded by at least five other 1's. This definition of environmental state takes into account how the organism responds to ('experiences?') the heterogeneity, and in this particular example it depends upon the fitness response parameter 'd'. Because this environment is not very runny, there is a chance of about 0.96 that any given generation will experience no 'bad' time units at all. Thus, we have a situation in which most generations experience only one environmental state ('good') while occasionally a generation will experience the other state ('bad'). This pattern of environmental heterogeneity could easily be classified as 'coarse-grained'.

We thus have two definitions of 'experience' leading to two different interpretations of grain. In the first case, we simply deal with the environmental states that occur without regard to the fitness response made by the individual to this heterogeneity. The judgment of how the individual 'experiences' the environments (fine or coarse-grained) is made by contrasting the time scale of environmental state transitions to the time scale of an individual's life. In the second case, 'experience' refers not only to relative time scales, but also to how the organism responds to the environments. Remember, we had to use the second definition of 'experience' when dealing with fitness set formulations. Since these definitions are originally due to Levins, perhaps the second, fitness-response, definition of 'experience' is the appropriate one. However, there are compelling reasons for avoiding this second definition.

The first reason is that the second definition makes the concept of grain unworkable in most practical situations. In order to decide the grain of an environment with this definition, one would not only have to observe the relative time scales of environmental state transitions to lifespan, but one would also have to perform laborious experiments designed to evaluate the organism's fitness response to the environmental states. This in practice has not been done by these authors discussing 'grain' in papers dealing with real organisms (e.g., Selander and Kaufman, 1973). In practice biologists use the first definition of 'experience' and avoid the second one which is consistent with fitness set formulations.

There are also theoretical reasons for avoiding the second definition of 'experience'. First, populations consist of genetically heterogeneous individuals, each with a potentially different fitness response. Consequently, the 'grain' of an environment can vary from genotype to genotype using the fitness-response definition of experience. For example, if $d=1$ instead of $d=5$ in Section 2, the grain would be 'fine' even with respect to the 'good' and 'bad' states. Note that this change of 'grain' occurs even though the time relation of the individual to the environmental states 0 and 1 are unchanged. Moreover, if the population consisted of some individuals with $d=1$ and others with $d=5$, the 'grain' would be fine for some and coarse for others, yet all individuals are members of the same evolving population, and the population can have only one evolutionary fate. With the second definition of experience, we can never speak of a population inhabiting a fine or coarse-grained environment because grain is now an attribute of an individual and not a population. This is indeed a serious flaw in the fitness-response definition of 'experience' since Levins and many other authors use grain at the population level. These difficulties vanish when one drops 'fitness-response' from the definition of 'experience'.

The second theoretical difficulty occurs because the second definition of experience implies fundamental evolutionary distinctions between situations which in fact are very similar. In fitness set analysis, grain determines the form of the adaptive function and hence governs the basic rules under which evolution (actually 'optimization') proceeds. Thus, one would expect to find major differences in the formulations of evolution under the fine versus coarse-grained environments. In this regard, let us return once again to the model of Section 2 and now allow the environmental state matrix to become increasingly runny for state 1. Under these conditions, it becomes more and more likely that every generation 'experiences' a mixture of both 'good' and 'bad' time units, rather than the coarse-grained pattern that occurred with respect to 'good' and 'bad' when using matrix (2-14). Yet all these cases would be governed by equations (2-8), (2-10), (2-12) and (2-13) and no fundamental shift in the form of the equations governing the evolutionary process occurs as in the case of changing grain in fitness set analysis. Rather, the equations referred to are justified by the relation of the time scale of environmental transitions of 0's and 1's to lifespan so that Good's (1973) formulae may be used; i.e., 'experience' in the first and more straightforward sense. Consequently, we feel that both practical and theoretical considerations argue against using the organism's response to environmental heterogeneity as a criterion for 'grain'. Instead, the word 'experience' in the definition of grain should simply refer to the relative time scales of environmental state transitions to individual lifespan regardless of individual fitness responses.

Even granting that 'experience' is defined as above, there are still ambiguities in the manner in which grain is used in evolutionary models. Fluctuations that occur between generations are generally classified as 'coarse-grained'. In the models developed in this paper, time stationarity was assumed so that the 'average' environment was identical every generation, but since each generation only made a finite sample of the within-generation fine-grained heterogeneity, fluctuations in realized environments would occur between generations due to sampling error. Does this mean that we are really dealing with a coarse-grained model? Alternatively, does this mean that any strictly fine-grained model must invoke no between-generation fluctuations whatsoever? The answers to these questions depend upon resolving additional ambiguities that are contained in the original definition of grain by Levins.

The sampling error in our model causes fitness to have a variance. This necessitates treating fitness as a random variable rather than as a constant. If there were no between-generation fluctuations in realized environment at all, the fitness would be a constant. (The fitness response is deterministic given a realized environment.) Consequently, the two questions stated above boil down to this: must fitness be treated as a random variable or as a constant for fine-grained environments? There is no clear statement on this matter in Levins (1968), although he did state that there is 'no uncertainty' in fine-grained environments. The implication is that fitness in a fine-grained environment may be treated as a constant. This idea was stated more explicitly by Strobeck (1975):

For any model of a strictly fine-grained environment, the conditions in each generation must be the same. Therefore the fitness values of the individuals of a given genotype will have the same distribution each generation, giving the genotype a constant fitness value.

Furthermore, from the same paper:

Because the individuals of a given genotype have the same distribution of fitness values every generation, the fitness value of the given genotype is a constant.

However, in all the models given in this paper the individuals of a given genotype also have the same distribution of fitness values every generation because time stationarity is assumed; yet, we treat fitness as a random variable and not as a constant. The discrepancy occurs because the justification of the constant fitness model for fine-grained environments depends upon much more than the assumption that every individual fitness is a sample from an identical distribution every generation. As we will now show, the justification of the constant fitness model depends upon the sampling properties of the whole population and not just the individual. This is an important point because almost all definitions of grain in the literature, including Levins', deal only with how the *individual* samples its environment, and not the population of individuals.

Strobeck's argument can be stated as follows. First, an individual's realized fitness enters the equations describing allele frequency changes only through its contribution to the average fitness of all individuals sharing that genotype. If V is the variance of individual fitness for a given genotype, N the population size and x the genotype frequency, then the variance of the average fitness is $\overline{V} = V/(Nx)$ if all individual fitnesses are *independent* realizations from the same distribution. If an infinite population size is assumed (as both Strobeck and ourselves are assuming), the $\overline{V} \rightarrow 0$ and a constant fitness model is appropriate. However, there are at least two sets of conditions under which this argument breaks down.

The first occurs when the assumption of independence of all individuals is not valid. This happens when the individuals within a generation experience their environment in a correlated fashion. For example, consider the evolution of a local population living in a small area. Let this population experience a succession of states of weather which influences fitness. Further, suppose the parameters determining the stochastic weather process are identical for all generations. On the average, the weather conditions are identical, but for any given generation all individuals experience exactly the same sequence of weather states. In this case, the variance of the average fitness equals the variance in individual fitness because all individuals share a single realization of the weather process, and not Nx independent realizations. This is the case we have treated in this paper, and as we have shown this causes fluctuations between generations and necessitates treating fitness as a random variable.

The second case occurs when Nx does not go to infinity even with independent sampling. This will occur whenever N is small, or when x is on the order of N^{-1} (which is always true for a newly arisen mutant) even if N is effectively infinite. The quantity x can also be rather small when dealing with multi-locus genotypes rather than single locus genotypes. Under these conditions the individual sampling differences in fitness are not 'averaged' out due to small within-generation sample size. This also leads to between generation fluctuations in realized average fitness. This case is treated in more detail in another paper (Templeton, 1977). This paper and the present one show that one must either be cautious about saying that fine-grained environments have 'no uncertainty' or change the definition of fine-grained away from the relation of the *individual* to *its* environment to include such population factors as the type of sampling scheme at the population level, the size of the population and the frequency of the individual's genotype in the total population. However, to include conditions on all these factors would once again make the concept of grain unworkable or, at the very least, extremely cumbersome. Hence, it is probably best to define grain solely in terms of the relationship of the individual to its environment and to recognize that fine-grained heterogeneity under many conditions can cause between-generation fluctuations.

In a very fundamental sense all of these semantic difficulties with grain are unnecessary. We could have written this entire paper and come up with identical results without once invoking the word 'grain'. Grain was simply used in this paper as a convenient summary description to categorize types of environmental heterogeneity. This is quite unlike the situation with fitness set analysis in which grain plays a critical role in determining the optimal population by influencing the nature of the adaptive function. The reason for this difference lies in the kind of question we have attempted to answer. As stated in the introduction we are concerned with the problem of *evolution* in heterogeneous, random environments and not *optimization* in the face of environmental uncertainty. Our basic approach to this evolutionary question has been to start with a stochastic model of environmental heterogeneity which is then coupled with a model of individual fitness response to secondarily generate a random fitness process. Genetic constraints are then overlayed upon the fitness process so that the evolutionary fate of a population may be determined. At times we need 'optimality' criteria, but these are defined in terms of the evolutionary fate of the population rather than defining the evolutionary fate in terms of the optimality criterion as in fitness set analysis.

This approach has many advantages. First, by treating the environment as our primary random variable, the models developed here are ecological models as well as genetic models. This allows us to study how a population of organisms can respond to various environmental patterns. We can also shift our emphasis to the organisms in order to study the evolution of various types of fitness responses given an environmental pattern. Models which treat fitness as the primary random variable cannot be used to investigate either of these problems. Second, by dealing with genetic constraints we can specify the evolutionary outcome given an environmental and fitness response model. There is no need to invoke adaptive functions, grain or other such *ad hoc* entities that are needed in optimization models. Considering the many ambiguities that are contained in such phrases as 'maximizing the adaptive function' or 'grain', this is a very valuable feature of our model. In this paper we have just begun to exploit the strengths of this approach. There are many directions in which we can expand this basic model by developing additional environmental models, fitness response functions, and genetic constraints.

By using this approach we have been able to make specific predictions about adaptive strategies without using decision theory. For example, equation (3-16) allows us to predict when the evolutionary process favors an individual that puts more energy into reducing its long-term detrimental responses to a prolonged, 'bad' environmental state as opposed to one that puts its energy into 'hiding' (in a fitness sense) completely from the environmental state for a short period of time. Similarly, equation (4-15) tells us when the strategic behavior of habitat selection will evolve as opposed to random dispersal. Finally, equation (5-10) tells us when females that choose to reproduce asexually will displace those reproducing sexually. The important point is that these 'decision rules' were not obtained by the application of decision theory to evolution, but rather were deduced from the properties of the evolutionary process itself. At times (e.g., Section 2), these criteria were not in accordance with the predictions made using a decision theoretic approach. When such a conflict exists, we feel that the decision theoretic prediction should be discarded because the population genetic approach provides not only the decision criterion, but the mechanism by which it is achieved in accordance with the natural laws that govern the process. We thus foresee a diminishing applicability of decision theory to evolution. However, the approach used in this paper did produce decision rules, and in that sense we have outlined a new type of decision theory (or perhaps added an extension onto the old). Perhaps the most important lesson to be learned from this work is that a 'decision' may already by build into a system that is sufficiently constrained by various natural laws and interactions between elements when the proper utility function is specified (in our case, the fitness functions). We have demonstrated the applicability of this principle to evolution, but it could easily be applicable to other areas. For example, a human makes his decisions not solely with a utility function, but operates in a system with many political, economic, and social constraints and in which interactions with, and opinions of, other individuals must be considered. This system of constraints coupled with the utility may well have built into it the strategic choice a person could make, with no need to assume *a priori* criteria, such as the Bayes or minimax principles. Thus, although decision theory may only have limited applicability to evolution, perhaps evolution has much applicability to decision theory.

Washington University, St. Louis, and University of Michigan

BIBLIOGRAPHY

Abramowitz, M. and Stegan, I. A. (eds.), *Handbook of Mathematical Functions*, Dover Publications, Inc. New York, 1965.

Adrewartha, H. G. and Birch, L. C., *The Distribution and Abundance of Animals*, University of Chicago Press, Chicago, 1954.

Asher, J. H., Jr., 'Parthenogenesis and Genetic Variability. II. One-locus Models for Various Diploid Populations', *Genetics* **66** (1970), 369–391.

Carson, H. L., 'Fixed Heterozygosity in a Parthogenetic Species of *Drosophila*', University of Texas Publication No. 6205, 1962, pp. 55–62.

Carson, H. L., 'Permanent Heterozygosity', *Evolutionary Biology*, **1** (1967), 143–168.

Emlen, J. M., *Ecology: An Evolutionary Approach*, Addison-Wesley Pub. Co., Reading, Massachusetts, 1973.

Felsenstein, J., 'The Evolutionary Advantage of Recombination', *Genetics* **78** (1974), 737–750.

Fisher, R. A., *The Genetical Theory of Natural Selection*, Oxford University Press, London, 1930.

Gabriel, K. R. and Neumann, J., 'A Markov Chain Model for Daily Rainfall Occurrence at Tel Aviv', *Quart. J. R. Met. Soc.* **88** (1962), 90–95.

Gillespie, J., 'The Role of Environmental Grain in the Maintenance of Genetic Variation', *American Naturalist* **108** (1974), 831–836.

Good, I. J. 'The Joint Probability Generating Function for Run-lengths in Regenerative Binary Markov Chains, with Applications', *Annals of Statistics* **1** (1973), 933–939.

Haldane, J. B. S., *The Causes of Evolution*, Longmans, Green, and Company, London, New York, 1932.

Hartl, O. L. and Cook, R. D., 'Stochastic Selection in Large and Small Populations', *Theoretical Population Biology* **7** (1975), 55–63.

Hunt, L. T. and Dayhoff, M. O., 'Table of Abnormal Human Globins', *Annals of New York Academy of Science* **241** (1974), 722–735.

Karlin, S. and Lieberman, U., 'Random Temporal Variation in Selection Intensities: Case of Large Population Size', *Theoretical Population Biology* **6** (1974), 355–382.

Koehn, R. K. and Mitton, J. B., 'Population Genetics of Marine Pelecypods I. Ecological Heterogeneity and Evolutionary Strategy at an Enzyme Locus', *American Naturalist* **106** (1972), 47–56.

Lerner, I. M., *Genetic Homeostasis*, Dover Publications Inc., New York, 1954.

Levins, R., 'Theory of Fitness in a Heterogeneous Environment. I. The Fitness Set and Adaptive Function', *American Naturalist* **96** (1962), 361–373.

Levins, R., *Evolutionary in Changing Environments*, Princeton University Press, Princeton, N.J., 1968.

Levins, R. and MacArthur, R., 'The Maintenance of Genetic Polymorphism in a Spatially Heterogeneous Environment: Variations on a theme by Howard Levens', *American Naturalist* **100** (1966), 585–589.

Lewontin, R. C., 'Studies on Heterozygosity and Homeostasis, II', *Evolution* **12** (1958), 494–503.

Lewontin, R. C., 'Is Nature Probable or Capricious?', *Bioscience* **16** (1966), 25–27.

Lewontin, R. C., *The Genetic Basis of Evolutionary Change*, Columbia University Press, New York, 1974.

Lokki, J., Suomalainen, E., Saura, A., and Lankinen, P., 'Genetic Polymorphism and Evolution in Parthenogenetic Animals. II. Diploid and Polyploid *Solenobia trignetrella* (Lepidoptera: Psychidae)', *Genetics* **79** (1975), 513–525.

Maynard Smith, J., 'What Use is Sex?', *Journal of Theoretical Biology* **30** (1971), 319–395.

Maynard Smith, J., 'Recombination and Rate of Evolution', *Genetics* **78** (1974), 299–304.

Murdy, W. H. and Carson, H. L., 'Parthenogenesis in *Drosophila mangabeirai* Malog', *American Naturalist* **93** (1959), 355–363.

Nyburg, D., 'Evolution of Sex. I. Primitive Sex', *Journal of Theoretical Biology* **50** (1975), 429–436.

Schwartz, D., 'An Example of Gene Fixation Resulting from Selective Advantage in Suboptimal Conditions', *American Naturalist* **103** (1969), 479–481.

Schwartz, D. and Laughner, W. J., 'A Molecular Basis for Heterosis', *Science* **166** (1969), 626–627.

Seiler, J., 'Untersuchungen über die Entstehung der Parthenogenese bei *Solenobia triquestrella* F. R. (Lepidoptera, Psychidae). III. Die geographische Verbreitung der drei Rassen von Solenobia triquetrella (bisexuell, diploid und tetraploid parthenogenetisch) in der Schweiz und in den angren zenden Ländern und die Beziehung zur Eiszeit. Bemerkungen uber die Entstehung der Parthenogenese', *Z. Vererb.* **92** (1961), 261–316.

Seiler, J., 'Untersach über die Entstehung der Parthenogenese bei *Solenobia triquetrella* F. R. (Lepidoptera, Psychidae). VII. Versuch einer experimentellen Analyse der Genetik der Parthenogenese', *Molecular and General Genetics* **99** (1967), 274–310.

Selander, R. K. and Kaufman, D. W., 'Genic Variability and Strategics of Adaptation in Animals', *Proc. Nat. Acad. Sci. U.S.A.* **70** (1973), 1875–1877.

Strobeck, C., 'Selection in a Fine-grained Environment', *American Naturalist* **109** (1975), 419–425.

Templeton, A. R., 'Density Dependent Selection in Parthenogenetic and Self-mating Populations', *Theoretical Population Biology* **5** (1974), 229–250.

Templeton, A. R., 'Survival Probabilities of Mutant Alleles in Fine-grained Environments'. American Naturalist III (1977), 951-966.

Templeton, A. R. and Rothman, E. D., 'The Population Genetics of Parthenogenetic Strains of *Drosophalia mercaturum*. I. One Locus Model and Statistics', *Theor. and Appl. Genetics* **43** (1973), 204–212.

Templeton, A. R., and Rothman, E. D., 'Evolution in Heterogeneous Environments', *American Naturalist* **108** (1974), 409–428.

Thoday, J. M., 'Balance, Heterozygosity and Developmental Stability', Cold Spring Harbor Symposia on Quantitative Biology **20** (1955), 318–326.

Vanzolini, P. E., 'Unisexual *Cnemidophoras lemmiseatus* in the Amazonas Valley: A Preliminary Note (Sauria, Teiidae)', *Papeis Avalsos de Zoologia* (Sao Paulo) **23** (1970), 63–68.

Williams, G. C., *Sex and Evolution*, Princeton University Press, Princeton, N.J., 1975.

Williams, G. C. and Mitton, J., 'Why Reproduce Sexually?' *Journal of Theoretical Biology* **39** (1973), 545–554.

Williamson, G. B., and Nelson, C. E., 'Fitness Set Analysis of Mimetic Adaptive Strategies', *American Naturalist* **106** (1972), 525–537.

Wright, S., 'The Roles of Mutation, Inbreeding, Crossbreeding, and Selection in Evolution', *Proceedings of the 6th International Congress of Genetics* **1** (1932), 356–366.

DONALD WITTMAN

POWER IN ELECTORAL GAMES

The study of collective decisions is the study of power. In order to know what decisions are made it is first necessary to know who has the power to make and implement them. 'Power' is a common sense term that one intuitively feels refers to major political and social phenomena. Procedures for collective decision making are required because there is not a consensus in the collectivity. Somehow, the power of the contenders seems to determine the distribution of wins and losses. Perhaps because the term has such a general reference, students of the subject have produced definitions without clearly specifying the scope of the definitions' applicability. The lack of communsurability of scope has made the comparison of the definitions difficult. Rather than banish the concept from scientific discourse (as proposed by William Riker),[1] this paper tries to clarify it by carefully defining the scope of this application and showing that, within that limited scope, application of several of the major definitions of power yields identical results.

This paper analyzes power exclusively in the context of N-person voting games, with two person games as a special case.[2] Analysis of definitions of power in a formal model of a political conflict enables one to draw deductive conclusions from the application of various definitions and thus determine unambiguously whether they are equivalent, contradictory, or in some other logical relation to each other. The results thus derived in this paper are not at all obvious from a mere semantic analysis of the definitions. The particular choice of N-person voting games was dictated by two considerations. First, in N-person voting games the players' only resources are votes, and the only possible outcomes are wins and losses (and ties, which we ignore here). This reduces the need to make questionable quantitative assumptions and simplifies the problem of calculating a measure of power.[3] Second, despite its relative mathematical simplicity, the theory of N-person voting games can be used to analyze non-trivial political issues, in particular constitutional ones, such as those raised in the debate between representatives of large and small states at the Constitutional Convention over the distribution of votes in the U.S. Congress.

Hooker, Leach, and McClennen (eds.), Foundations and Applications of Decision Theory, Vol. II, 185–206. *All Rights Reserved*

1. Power in Two Person Voting Games: Probability of a Bill Passing

The two person case will be considered first. Later it will be subsumed under the formula for N-person games. However, to analyze power in N-person games it is necessary to calculate the probability of various coalitions forming. In the following two person game-like situation these complex calculations can be avoided and thereby the analysis of power can be presented more directly. It should then be easier for the reader to follow the analysis of N-person games.

To make the analysis less abstract, power will be considered within the context of passing a legislative bill. Each player has two strategies: to be for or against the bill. The two players are not the only voters, but the other voters are not treated as game players. Their votes are some probability of the two players' choices of strategies. One may conceive of the two players as leaders of factions, with some stochastic control over the votes of their followers. The probability of the bill passing is thus a function of the strategy choices of the two players, and can be put into a two by two matrix (see Table 1). Note that the entries in this matrix are neither cardinal nor ordinal utilities, as in the normal form of a game, but descriptions of states of the world over which the players' utility or preference functions would be defined. This matrix is the object analyzed in the study of power in two person voting games.

The first definition of power I will look at is the one originally proposed by Robert Dahl and used by him and his collaborators to analyze the influence of U.S. Senators.[4] I call this the player reversal approach to power. This approach ranks the two players ordinally by examining the outcomes in cases where the players' goals are opposed. One compares the probability of a bill passing when Player 1 is for and Player 2 is against with the probability of its passing when the positions are reversed. In terms of our matrix, one compares P_{12} with P_{21}. If $P_{12} > P_{21}$, then Player 1 is more powerful than Player 2; and the reverse if $P_{12} < P_{21}$. In the hypothetical numerical example given in Table 1, since $70 > 60$, player 1 is more powerful. There is a rationale for looking only at the cases where the two players are opposed. When they agree on the outcome their relative power is not tested. As in a football game, where we gauge the skill of the teams by having them take turns scoring against each other, we observe the two players in cases where they are trying to defeat each other. In cases where they agree we have no way of knowing how much each contributes to the outcome. This lack of

TABLE 1

Probability of the bill passing

		Player 2	
		For	Against
Player 1	For	90	70
	Against	60	20
		Player 2	
		For	Against
Player 1	For	P_{11}	P_{12}
	Against	P_{21}	P_{22}

Player 1 is said to be more powerful than Player 2.

A. Player reversal $P_{12} > P_{21}$.
B. Range of possibilities $P_{12} - P_{21} > P_{21} - P_{12}$.
C. Restricted outcome $P_{11} - P_{12} < P_{11} - P_{21}$ and $P_{21} - P_{22} < P_{12} - P_{22}$.
D. Ability to harm $P_{11} - P_{12} < P_{11} - P_{21}$ and $P_{21} - P_{22} < P_{12} - P_{22}$.
E. Path analysis $P_{12} - P_{22} > P_{21} - P_{22}$.

independent information is analogous to the identification problem in multivariate statistics, as will become clear in the discussion below of the path analysis approach to power by Nagel.

The second approach derives from Weber's definition of power: '*Macht* is the probability that one actor within a social relationship will be in a position to carry out his own will despite resistance'.[5] This approach also focuses on the cases where the players are opposed, i.e., when there is resistance. But rather than looking simply at the probabilities of the bill's passing, this approach looks at the range of probabilities a player can guarantee despite opposition from his antagonist. To measure Player 1's power we look at the difference between the least probability of the bill passing when he is for and the greatest probability of its passing when he is against. If this is greater than the analogous difference for 2, we say he is the more powerful player. In terms of our matrix the condition for 1 being more powerful than 2 is:

$$\min_j P_{1j} - \max_j P_{2j} > \min_i P_{i1} - \max_i P_{i2}.$$

I now make the rather modest assumption that either player's being for the bill increases its chances of passing and either player's being against it decreases its chances (holding the other player's choice constant). This assumption holds throughout the rest of the paper. With this assumption, the above formula becomes: $P_{12} - P_{21} > P_{21} - P_{12}$. In our hypothetical case: $70 - 60 > 60 - 70$, so Player 1 is more powerful.

James March, as well as J. Thibaut and H. Kelley, have suggested a third criterion for determining the power of social actors, that of the ability to restrict outcomes.[6] A powerful actor will be able to restrict the deviations of outcomes from his preferred outcome more effectively than a weak actor, whatever his preference may be. In this case, since there are two choices, there are two conditions sufficient to state that Player 1 is more powerful than Player 2. When Player 1 is for, he must be able to restrict outcomes more than 2 can when 2 is for; and similarly when he is against. In our symbols, the conditions are: $P_{11} - P_{12} < P_{11} - P_{21}$ and $P_{21} - P_{22} < P_{12} - P_{22}$. Once again, Player 1 is more powerful: $90 - 70 < 90 - 60$ and $60 - 20 < 70 - 20$.

The fourth approach was developed by George Karlsson from the behavioral study of small groups and is based on the ability to harm.[7] The more harm one player can do to another, the more power he has over him. If we assume that players in our voting game choose strategies to coincide with preferred outcomes, then the harm one player can do another is the amount by which he can diminish the probability of the other's desired outcome. Similarly, the good he can do is the amount by which he can augment the probability of the desired outcome. Since the two quantities are the same, we suffer no loss of generality by looking only at the possible harm. As in the third approach, there result two conditions for Player 1 being more powerful than Player 2. The amount of harm 1 can do to 2 if 2 is for must be greater than the amount of harm 2 can do to 1 if 1 is for; and similarly when they are against. If 2 is for, 1 can choose between the values in the first column, so the amount of harm is $P_{11} - P_{12}$; analogously for the other case. Conditions for 1 being more powerful than 2 are: $P_{11} - P_{21} > P_{11} - P_{12}$ and $P_{12} - P_{22} > P_{21} - P_{22}$. In our case: $90 - 70 > 90 - 60$ and $70 - 20 > 60 - 20$ so 1 is more powerful than 2.

The last approach considered here is one developed by Jack Nagel, which uses the statistical methods of path analysis to measure the marginal effects of actors upon outcomes.[8] To apply his method to our case we set up the following regression equation: $P_{ij} = A + B_1X_1 + B_2X_2 + B_3X_1X_2$, where X_i is a dummy variable: $X_i = 1$ if i is for, $X_i = 0$ if i is against. (Nagel usually ignores the interaction term.) B_1 and B_2 can be estimated to give the marginal effect, here equated with power, of Players 1 and 2 on the outcome. If $B_1 > B_2$, Player 1 is more powerful. Before solving the regression equation let me point out that this equation shows, as was noted in discussing the player reversal approach, that observations of cases where the players agree do not enable us to estimate their power. When they agree, $X_1 = X_2 = X_1 X_2$; the supposedly independent variables are linearly dependent and the parameters are underidentified. Substituting the values of X_i gives the following equations:

$$\begin{aligned} P_{11} &= A + B_1 + B_2 + B_3 \\ P_{12} &= A + B_1 \\ P_{21} &= A + B_2 \\ P_{22} &= A. \end{aligned}$$

The solutions are:

$$\begin{aligned} A &= P_{22} \\ B_1 &= P_{12} - P_{22} \\ B_2 &= P_{21} - P_{22} \\ B_3 &= P_{11} - P_{12} - P_{21} + P_{22}. \end{aligned}$$

The condition for 1 being more powerful than 2 is $P_{12} - P_{22} > P_{21} - P_{22}$. With our numbers: $P_{ij} = 20 + 50X_1 + 40X_2 - 20X_1X_2$. Player 1 is more powerful.

Table 1 lists the criteria for Player 1 being more powerful than Player 2 given by all five definitions of power. Each of these formulae can be derived from every other formula, as a little algebraic manipulation will convince the reader.[9] Since each formula can be reduced to that given by the Player Reversal approach, which includes only the cases where the actors are opposed, none of the formulas in Table 1 requires observations of cases where the actors agree to determine the power ranking of the players (although such observations are required to obtain cardinal results by Nagel's method). This is what the brief discussion at the beginning of the paper suggested: power is a concept used in describing situations where conflict requires the imposition of a binding rule, not those where a consensus exists.

2. A PRIORI VOTING POWER IN N-PERSON ELECTORAL GAMES

In the previous section the probabilities entered in the matrix were assumed to be determined *a posteriori*, by observation. To analyze N-person games we will examine various methods for calculating the probability of a player's belonging to a winning coalition *a priori*, that is, only on the basis of knowledge of the voting rules in the game. To do so we must determine the probability of formation of each possible coalition. Different researchers have proposed different ways of calculating these probabilities. L. S. Shapley proposed that every ordering of a coalition is equally likely.[10] J. Banzhalf and J. Coleman suggested that every coalition is equally likely.[11] Coleman also explored a model in which each person has probability *P* of being in a coalition.[12] W. Riker and his followers presumed that every minimal winning coalition is equally likely.[13] These various principles for predicting coalition formation could attach different probabilities to the possible coalitions and lead to different power rankings of the players even under the same definition of power. If slightly different plausible assumptions yield such different results we would have little confidence in one result. In the remainder of this paper I present cases of voting games where the different formulas tend to produce similar results.

The games studied here are all two stage voting games, also called compound-simple games. In these games there is a set of voters divided into groups. In the first stage votes are taken within distinct (non-overlapping) groups. In the second stage the groups vote as units, according to the decision of their members in the first stage. Clearly this type of voting game typifies many real electoral institutions.

One may divide two stage voting games into three pure types. In Type I each group has the same number of voters in the second stage regardless of the number of its members. Such a scheme has recently been proposed in Lebanon, where the parliamentary seats are apportioned among Christians and Muslims without regard to size of population. One proposed scheme is to give each community half the votes: a pure Type I system. The U.S. Senate is also a pure Type I voting game.

In Type II the groups have the same number of members but different numbers of votes. The most familiar example of such a scheme is the U.N. Security Council, whose requirement that all five permanent members and two of the six temporary members assent to a resolution is isomorphic to a system which gives six votes to each of the former and one vote to each of the latter, while requiring thirty-two votes for passage.[14] This is a degenerate type of two stage game, however, since each group has only one member.

The structures of Type I and Type II voting games are such that reasonable ordinal rankings of power can be derived without recourse to sophisticated mathematical analysis. In Type I games, the members of smaller groups never have less power than the members of large groups, since the former make a larger proportional contribution to their group's single vote. Similarly in Type II games, the member of a group with more votes cannot have less power than the members of a group with fewer votes since the former's equally weighted first stage votes are magnified in effect on the final outcome by the unequal weights at the second stage.

Type III games are the hardest to analyze. Institutions of this form were generally designed to give voters approximately equal power, and there is no commonsensical way to determine how the voting procedures are in fact biased. In these games the groups have different numbers of members, and the number of votes a group has is proportionate to the number of its members. The most commonly cited example of such a system is the U.S. Electoral College. Arguments have been made that both large state voters and small state voters are more powerful under the system, and there seems to be no way for common sense argument to settle the question. The method of analysis proposed here does settle it. All the formulae lead to identical conclusions. The power relationships in Type III games depend in a well defined way on the particular configuration of sizes of voting groups. In some configurations the small group voters are more powerful, while in others the large group voters are more powerful.

I can illustrate this result with numerical examples before going on to prove it in general. Suppose there are three groups of voters: *A, B,* and *C*. *A* and *B* have three members each; *C* has one member. Since the game is a Type III game the votes of the groups are exactly proportionate to the number of members, i.e., the votes are some multiple of the triple (3, 3, 1). If a majority is needed to win, there are four winning coalitions: (*A, B*) (*A, C*), (*B, C*) or (*A, B, C*). Under the hypothesis of equiprobability of coalitions, each group is a member of three of four winning coalitions. Thus each group is equally powerful and the member of the small group is more powerful than the members of the large group, since he has complete control over whether his group joins the coalition. The other formulae give the same result, or a stronger one. Incidentally, this game is isomorphic to a Type I voting game in which each group has one vote, in the sense that the set of winning coalitions is identical.[15] Clearly in such a game the members of the small group are more powerful.

A game which yields the opposite result is one in which, of the three groups, one has three voters and the other two have one each. The large group is always in the winning coalition, while the other groups may not be. Indeed, since the large group itself constitutes a majority, the small groups have no effect whatever on the outcome and hence no power.

These are the polar possible results. I now turn to a more general case in which there is a single large group with $N+1$ voters and $L+1$ small groups with one voter each. I will analyze this case with each of the varying methods of computing *a priori* probabilities of coalition formation. The methods all yield identical ordinal power rankings of the voters. That is, if voter A is more powerful than voter B according to one formula, he is more powerful than voter B according to the other formula as well.

METHOD I. All voters except A, who is in the large voting unit, and B, who is in one of the small voting units, vote for X with probability P. I will now show that for all values of P (between 0 and 1) when A votes for and B votes against, the probability of X winning is greater than when A votes against and B votes for.

Proof. For $N+1 > L+1$ the proof is obvious as the voting unit with $N+1$ voters has a majority of votes and thus the $L+1$ voting units have no effect on the outcome. Henceforth I will assume $N < L$. N even, L odd.

(1) Let $S =$ the probability that X wins when A votes for and B votes against.

(2) Let $T =$ the probability that X wins when B votes for and A votes against.

To prove $S > T$:

(3)
$$S = \left[\sum_{j=\frac{N}{2}}^{N} \binom{N}{j} P^j Q^{N-j}\right] \left[\sum_{\substack{k = \frac{N+1+L+1+1}{2} - N - 1 \\ = \frac{-N+L+1}{2}}}^{L} \binom{L}{k} P^k Q^{L-k}\right]$$

$$+ \left[1 - \sum_{j=\frac{N}{2}}^{N} \binom{N}{j} P^j Q^{N-j}\right] \left[\sum_{i=\frac{N+1+L+1+1}{2}}^{L} \binom{L}{i} P^i Q^{L-i}\right].$$

I.e., in words, the probability that the group votes for X when A votes for and B votes against =

$\left[\text{The probability that the voting unit with } N+1 \text{ voters votes for. I.e., the probability that at least } \frac{(N+1)}{2}-1 \text{ voters besides } A \text{ vote for.}\right]$ times $\left[\text{The probability that at least enough more voters not in the voting unit with } N+1 \text{ voters vote for } X \text{ so that } X \text{ wins the election. } \frac{N+1+L+1+1}{2} \text{ is a majority so this is the probability that } \frac{(N+1+L+1+1)}{2}-(N+1) \text{ voters in the single voter units vote for } X.\right]$

$+\left[\text{The probability that the voting unit with } N+1 \text{ voters votes against } X.\right]$ times $\left[\text{The probability that at least } \frac{(N+1+L+1+1)}{2} \text{ voters in the single voter units vote for } X \text{ so that } X \text{ wins the election.}\right]$

(4)

$$T = \left[\sum_{j=\frac{N}{2}+1}^{N} \binom{N}{j} P^j Q^{N-j}\right]\left[\sum_{k=\frac{N+1+L+1+1}{2}-(N+1)-1 = \frac{-N+L-1}{2}}^{L} \binom{L}{k} P^k Q^{L-k}\right]$$

(the final -1 in the lower limit: B already voting for X)

$$+ \left[1-\sum_{j=\frac{N}{2}+1}^{N} \binom{N}{j} P^j Q^{N-j}\right]\left[\sum_{i=\frac{N+1+L+1+1}{2}-1 = \frac{N+L+1}{2}} \binom{L}{i} P^i Q^{L-i}\right]$$

(the final -1 in the lower limit: B already voting for X.)

Breaking up S into parts:

(5)

$$S = \binom{N}{\frac{N}{2}} P^{\frac{N}{2}} Q^{\frac{N}{2}} \sum_{k=\frac{-N+L+1}{2}}^{\frac{N+L-1}{2}} \binom{L}{k} P^k Q^{L-k}$$

$$+ \underbrace{\binom{N}{\frac{N}{2}} P^{\frac{N}{2}} Q^{\frac{N}{2}} \binom{L}{\frac{N+L+1}{2}} P^{\frac{N+L+1}{2}} Q^{\frac{L-N-1}{2}}}_{a}$$

$$+ \underbrace{\binom{N}{\frac{N}{2}} P^{\frac{N}{2}} Q^{\frac{N}{2}} \sum_{k=\frac{N+L+3}{2}}^{L} \binom{L}{k} P^k Q^{L-k}}_{b}$$

$$+ \underbrace{\sum_{j=\frac{N}{2}+1}^{N} \binom{N}{j} P^j Q^{N-j} \sum_{k=\frac{-N+L+1}{2}}^{L} \binom{L}{k} P^k Q^{L-k}}_{c}$$

$$\underbrace{-\binom{N}{\frac{N}{2}} P^{\frac{N}{2}} Q^{\frac{N}{2}} \sum_{i=\frac{N+L+3}{2}}^{L} \binom{L}{i} P^i Q^{L-i}}_{-b}$$

$$+ \underbrace{\sum_{j=0}^{\frac{N}{2}} \binom{N}{j} P^j Q^{N-j} \sum_{i=\frac{N+L+3}{2}}^{L} \binom{L}{i} P^i Q^{L-i}}_{d}$$

Breaking up T into parts:

(6)

$$T = \sum_{j=\frac{N}{2}+1}^{N} \binom{N}{j} P^j Q^{N-j} \binom{L}{\frac{-N+L-1}{2}} P^{\frac{-N+L-1}{2}} Q^{\frac{N+L+1}{2}}$$

$$+ \underbrace{\sum_{j=\frac{N}{2}+1}^{N} \binom{N}{j} P^j Q^{N-j} \sum_{k=\frac{-N+L+1}{2}}^{L} \binom{L}{k} P^k Q^{L-k}}_{c}$$

$$+ \underbrace{\binom{L}{\frac{N+L+1}{2}} P^{\frac{N+L+1}{2}} Q^{\frac{L-N-1}{2}} \sum_{j=0}^{\frac{N}{2}-1} \binom{N}{j} P^j Q^{N-j} + \binom{L}{\frac{N+L+1}{2}} P^{\frac{N+L+1}{2}} Q^{\frac{L-N-1}{2}} \binom{N}{\frac{N}{2}} P^{\frac{N}{2}} Q^{\frac{N}{2}}}_{a}$$

$$+ \underbrace{\sum_{k=\frac{N+L+3}{2}}^{L} \binom{L}{k} P^k Q^{L-k} \sum_{j=0}^{\frac{N}{2}} \binom{N}{j} P^j Q^{N-j}.}_{d}$$

Both S and T have elements in common. Subtracting off common elements a, c and d we have: $S^* = S-a-c-d$ and $T^* = T-a-c-d$. For A:

(7)

$$S^* = \binom{N}{\frac{N}{2}} P^{\frac{N}{2}} Q^{\frac{N}{2}} \sum_{k=\frac{-N+L+1}{2}}^{\frac{N+L-1}{2}} \binom{L}{k} P^k Q^{L-k}.$$

S^* = The probability that A's vote is crucial for the group voting for when B is against.

For Person B:

(8)
$$T^* = \sum_{j=\frac{N}{2}+1}^{N} \binom{N}{j} P^j Q^{N-j} \binom{L}{\frac{-N+L-1}{2}} P^{\frac{-N+L-1}{2}} Q^{\frac{N+L+1}{2}}$$
$$+ \binom{L}{\frac{N+L+1}{2}} P^{\frac{N+L+1}{2}} Q^{\frac{L-N-1}{2}} \sum_{j=0}^{\frac{N}{2}-1} \binom{N}{j} P^j Q^{N-j}.$$

T^* = The probability that B's vote is crucial for the group voting for when A is against.

(9)
$$\overset{?}{\lessgtr} \binom{N}{\frac{N}{2}} P^{\frac{N}{2}} Q^{\frac{N}{2}} \sum_{i=\frac{-N+L+1}{2}}^{\frac{N+L-1}{2}} \binom{L}{i} P^i Q^{L-i} = S^*.$$

Each side has N expressions

For T^* $\quad \sum_{j=\frac{N}{2}+1}^{N} (\ ) + \sum_{j=0}^{\frac{N}{2}-1} (\ )$ has N parts to the summation.

For S^* $\quad \sum_{k=\frac{-N+L+1}{2}}^{\frac{N+L-1}{2}}$ means the number of parts

$$= 1 + \left(\frac{N+L-1}{2}\right) - \left(\frac{-N+L+1}{2}\right) = N \text{ parts to the summation}$$

↑
because includes first and last.

Each side has N expressions when multiplied out. Each of these expressions is smaller on the left hand side than on the right.

(10)
$$T^* = \left[\frac{N!}{\frac{N+2}{2}!\,\frac{N-2}{2}!} P^{\frac{N+2}{2}} Q^{\frac{N-2}{2}} + \frac{N!}{\frac{N+4}{2}!\,\frac{N-4}{2}!} P^{\frac{N+4}{2}} Q^{\frac{N-4}{2}} + \ldots + \frac{N!}{\frac{N+N}{2}!\,\frac{N-N}{2}!} P^{\frac{N+N}{2}} Q^{\frac{N-N}{2}}\right]\left[\frac{L!}{\frac{-N+L-1}{2}!\,\frac{N+L+1}{2}!} P^{\frac{-N+L-1}{2}} Q^{\frac{N+L+1}{2}}\right]$$

$$+\left[\frac{N!\,P^0 Q^N}{0!\,N!} + \frac{N!\,P^1 Q^{N-1}}{1!\,(N-1)!} + \ldots + \frac{N!}{\frac{N-2}{2}!\,\frac{N+2}{2}!} P^{\frac{N-2}{2}} Q^{\frac{N+2}{2}}\right]$$

$$\left[\frac{L!}{\frac{N+L+1}{2}!\,\frac{L-N-1}{2}!} P^{\frac{N+L+1}{2}} Q^{\frac{L-N-1}{2}}\right]$$

(11)
$$\overset{?}{\leq} \frac{N!}{\frac{N}{2}!\,\frac{N}{2}!} P^{\frac{N}{2}} Q^{\frac{N}{2}} \left[\frac{L!}{\left(\frac{-N+L+1}{2}\right)!\,\left(\frac{N+L-1}{2}\right)!} P^{\frac{-N+L+1}{2}} Q^{\frac{N+L-1}{2}} + \frac{L!}{\frac{-N+L+3}{2}!\left[L-\frac{-N+L+3}{2}\right]!} P^{\frac{-N+L+3}{2}} Q^{\frac{N+L-3}{2}} \ldots + \frac{L!}{\left(\frac{N+L-1}{2}\right)!\,\left(\frac{L-N+1}{2}\right)!} P^{\frac{N+L-1}{2}} Q^{\frac{-N+L+1}{2}}\right] = S^*.$$

Looking just at P, Q, if the ith term in eq. 10 has $P^X Q^Y$ then the ith term in eq. 11 has $P^X Q^Y$.

Proof. The first term in eq. 10 has

$$P^{\frac{N+2}{2}} Q^{\frac{N-2}{2}} P^{\frac{-N+L-1}{2}} Q^{\frac{N+L+1}{2}} = P^{\frac{L+1}{2}} Q^{\frac{2N+L-1}{2}} \tag{12}$$

$$= \text{First term in eq. 11} =$$

$$P^{\frac{N}{2}} Q^{\frac{N}{2}} P^{\frac{-N+L+1}{2}} Q^{\frac{N+L-1}{2}} = P^{\frac{L+1}{2}} Q^{\frac{2N+L-1}{2}}. \tag{13}$$

Each following term has the power of P raised by 1 and the power of Q lowered by 1.

I will now look at each term in T^* and show it to be $\leq$ to the corresponding term in S^*. By the above proof the P's and Q's can be ignored.

Looking at the first $N/2$ terms is

$$T_i^* = \binom{N}{\frac{N}{2}+i}\binom{L}{\frac{L+N+1}{2}} \leqslant \binom{N}{\frac{N}{2}}\binom{L}{\frac{L+N+1}{2}-i} = S_i^* \tag{14}$$

$$i = 1, 2, \ldots, \frac{N}{2}.$$

The prime terms in the denominator of T_i^* are close to N and 0 and L and 0 while the prime terms in the denominator of S_i^* are closer to $N/2$ and $L/2$; therefore, the denominator of S_i^* is smaller and consequently S_i^* is larger.

Proof:

$$\frac{S_i^*}{T_i^*} = \frac{\left(\frac{N}{2}+i\right)\left(\frac{N}{2}+i-1\right)\ldots\left(\frac{N}{2}+1\right)}{\left(\frac{N}{2}\right)\left(\frac{N}{2}-1\right)\left(\frac{N}{2}-i+1\right)} \tag{15}$$

$$\frac{\left(\frac{L+N+1}{2}\right)\left(\frac{L+N+1}{2}-1\right)\ldots\left(\frac{L+N+1}{2}-i+1\right)}{\left(\frac{L-N-1}{2}+i\right)\left(\frac{L-N-1}{2}+i-1\right)\ldots\left(\frac{L-N-1}{2}+1\right)} \geqslant 1$$

as each term in the numerator is $\geq$ each term directly below in the denominator.

(16) $\frac{N}{2}+i>\frac{N}{2}:\frac{L+N+1}{2}\geqslant\frac{L-N-1}{2}+i$ for $i=1,2,\ldots,\frac{N}{2}$.

Looking at the next $N/2$ terms,

(17) $$T_i^* = \binom{N}{i-1}\binom{L}{\frac{N+L+1}{2}} \leqslant \binom{N}{\frac{N}{2}}\binom{L}{\frac{L-1}{2}+i} = S_i^* \qquad i = 1, 2, \ldots, \frac{N}{2}.$$

As the terms in the denominator in S_i^* are closer to $N/2$ and $L/2$ than the terms in the denominator in T_i^*.

(18) $\therefore S^* > T^*$ and thus A is more powerful than B.

(Q.E.D.)

When P is either very high or very low (i.e., there is a consensus either for or against) the power of any participant is greatly reduced for he is unlikely to have an effect on the outcome.

METHOD II. All coalitions such that A is in one coalition and B is in the other are equally likely. This is identical to the approach used by Coleman and Banzhaf except that coalitions which include both B and A are not considered.

(19) Let S' be the probability that the group votes for when A votes for

(20) $$S' = \frac{\displaystyle\sum_{k=\frac{N}{2}}^{N}\binom{N}{k}\sum_{k=\frac{L+1-N}{2}}^{L}\binom{L}{k} + \sum_{k=0}^{\frac{N}{2}-1}\binom{N}{k}\sum_{k=\frac{N+L+3}{2}}^{L}\binom{L}{k}}{2^{N+L}}.$$

The numerator equals the number of coalitions where the group votes for. The denominator equals the number of possible coalitions. By Method 1, $S > T$ for all P. For $P = \frac{1}{2}$

$$(21)\qquad S = \sum_{k=\frac{N}{2}}^{N} \binom{N}{k}\left(\frac{1}{2}\right)^{N} \sum_{k=\frac{L+1-N}{2}}^{L} \binom{L}{k}\left(\frac{1}{2}\right)^{L} + \sum_{k=0}^{\frac{N}{2}-1} \binom{N}{k}\left(\frac{1}{2}\right)^{N} \sum_{k=\frac{L+N+3}{2}}^{L} \binom{L}{k}\left(\frac{1}{2}\right)^{L} = S'$$

(22) Similarly $T = T'$ for $P = \frac{1}{2}$. Therefore $S' > T'$.

METHOD III. A is said to be more powerful than B if the number of orderings of coalitions where A is the pivotal member of the winning coalition and B is the first member of the losing coalition (S'') is greater than the number of orderings of coalitions where B is the pivotal member of the winning coalition and A is the first member of the losing coalition (T''); i.e., ignoring A and B all orderings of coalitions are equally likely. The pivot position is that sequential position which turns the coalition from a losing coalition to a winning one. This is identical to the Shapley-Shubik formula except that $B(A)$ does not vary freely when $A(B)$ is the pivot.

$$(23)\qquad S'' = \sum_{k=\frac{L-N+1}{2}}^{\frac{N+L+1}{2}} \frac{\left(\frac{N}{2}+k\right)!}{\frac{N}{2}!} \, \frac{\left(\frac{N}{2}+L-k\right)!}{\frac{N}{2}!}.$$

$[(N/2) + k]!$ is the number of ways of ordering the members of the winning coalition before pivot A where each voter is identified. This is divided by $N/2!$, the number of ways of ordering the $N/2$ member of the voting unit with N voters who vote for, so that the expression does not distinguish voters within the voting unit. $[(N/2) + L - k]!$ is the number of ways of ordering the members of the losing coalition after B where each voter is identified.

$$(24) \qquad T'' = \sum_{k=\frac{N}{2}+1}^{N} \frac{\left(k + \frac{L-N-1}{2}\right)!}{k!} \frac{\left(N - k + \frac{L+N+1}{2}\right)!}{(N-k)!}$$

$$+ \sum_{k=0}^{\frac{N}{2}} \frac{\left(k + \frac{N+L+1}{2}\right)!}{k!} \frac{\left(N - k + \frac{L-N-1}{2}\right)!}{(N-k)!}$$

The first term refers to cases where the group composed of $N+1$ voters votes for. For B to be the pivot in such cases there must be one less than a majority voting for before B. Since $N+1$ votes are for, $[(N+1+L+1+1)/2] - (N+1) - 1$ additional voters are needed ahead of B. $[(N+1+L+1+1)/2] - N - 1 - 1 = (L-N-1)/2$. k refers to the number of people voting for in the $N+1$ group. If the $N+1$ group is voting for $(N/2)+1$ members of the group must be voting for. When B votes for, then A votes against. The remaining voters can then be ordered in any way. The number of voters voting against (not including A, B) = the total number of voters minus the number of voters ahead of B voting for $= N+1+L+1+1-[k+(L-N-1)/2] = N-k+[(L+N+1)/2]$. The second term considers cases where the $N+1$ group votes against ($k = 0, 1 \ldots N/2$). Then $[(N+1+L+1+1)/2] - 1 = (N+L+1)/2$ is the number of individuals not in group $N+1$ who vote for before B the pivot.

(25) Each element in S'' =

$$\frac{\left(\frac{L-1}{2} + i\right)!}{\frac{N}{2}!} \frac{\left(N + \frac{L+1}{2} - i\right)!}{\frac{N}{2}!} \qquad i = 1, 2, \ldots, N+1.$$

The first $N/2$ elements in T'' =

$$(26) \qquad \frac{\left(i + \frac{L-1}{2}\right)!}{\left(\frac{N}{2} + i\right)!} \frac{\left(N + \frac{L+1}{2} - i\right)!}{\left(\frac{N}{2} - i\right)!} \qquad i = 1, 2, \ldots, \frac{N}{2}.$$

The last $N/2 + 1$ elements in T'' =

$$(27) \qquad \frac{\left(i + \frac{N+L-1}{2}\right)!}{(i-1)!} \frac{\left(\frac{N+L+1}{2} - i\right)!}{(N-i+1)!} \qquad i = 1, 2, \cdots, \frac{N}{2} + 1$$

Looking at the first $N/2$ elements, the numerator is the same for corresponding elements in S'' and T''.

(28) The denominator in $S'' = \frac{N}{2}!\frac{N}{2}! \leqslant \left[\frac{N}{2}+i\right]!\left[\frac{N}{2}-i\right]!$

Therefore the first $N/2$ elements in S'' are larger than the first $N/2$ elements in T''.

The $N/2 + 1$ element in S'' =

(29) $$\frac{\left(\frac{L+1+N}{2}\right)!}{\frac{N}{2}!}\frac{\left(\frac{N+L-1}{2}\right)!}{\frac{N}{2}!}$$

The numerator is the same as the $(N/2) + 1$ element in T'' (i.e., the first element in the last $(N/2) + 1$ elements in T'') and again the denominator is smaller than the corresponding one in T''. This holds for each of the remaining elements, therefore $S'' > T''$. The similarity between S and S'' and T and T'' should not be overlooked.

Many authors believe that only minimal winning coalitions will take place. This method will yield the same results as the previous methods.

METHOD IV. All minimal coalitions such that A is in one coalition and B is in the other are equally likely.

(30) Let S''' be the number of minimal winning coalitions in which A is a member.

(31) $$S''' = \binom{N}{\frac{N}{2}}\binom{L}{\frac{N+L+3}{2}-N-1} = \binom{N}{\frac{N}{2}}\binom{L}{\frac{L-N+1}{2}}$$

$$= \frac{N!}{\frac{N}{2}!\frac{N}{2}!}\frac{L!}{\frac{L-N+1}{2}!\frac{L+N-1}{2}!}$$

(32) $$\overset{?}{\gtrless}\binom{N}{\frac{N}{2}+1}\left(\binom{L}{\frac{L-N-1}{2}}+\binom{L}{\frac{N+L+1}{2}}\right)$$

$$= \frac{N!}{\left(\frac{N}{2}+1\right)!\left(\frac{N}{2}-1\right)!}\frac{L!}{\frac{L-N-1}{2}!\frac{L+N+1}{2}!} + \frac{L!}{\frac{N+L+1}{2}!\frac{L-N-1}{2}!}$$

$= T'''$ the number of minimal winning coalitions in which B is a member.

$$(33)\qquad \frac{T'''}{S'''} = \frac{\frac{N}{2}\,\frac{L-N+1}{2}}{\left(\frac{N}{2}+1\right)\,\frac{L+N+1}{2}} + \frac{\frac{N}{2}!\,\frac{N}{2}!\,\frac{L-N+1}{2}}{N!\,\frac{L+N+1}{2}} \overset{?}{\leqslant} \frac{S'''}{S'''} = 1$$

$$(34)\qquad \frac{T'''}{S'''} = \underbrace{\frac{L-N+1}{L+N+1}}_{A}\underbrace{\left(\frac{N}{N+2}+\frac{\frac{N}{2}!\,\frac{N}{2}!}{N!}\right)}_{B}$$

$A < 1$, therefore is $B \leqslant 1$ or equivalenty

$$(35)\qquad \text{is}\ \frac{\frac{N}{2}!\,\frac{N}{2}!}{N!} \leqslant 1 - \frac{N}{N+2} = \frac{N+2-N}{N+2} = \frac{2}{N+2} = \frac{1}{\frac{N}{2}+1}.$$

Equivalenty is

$$(36)\qquad 1 \geqslant \frac{\left(\frac{N}{2}+1\right)}{N!}\,\frac{N}{2}!\,\frac{N}{2}! = \frac{\left(\frac{N}{2}+1\right)!\,\frac{N}{2}!\,\left(N+1\right)}{(N+1)!} = \frac{N+1}{\frac{(N+1)!}{\left(\frac{N}{2}+1\right)!\,\frac{N}{2}!}}$$

$$(37)\qquad = \frac{\binom{N+1}{N}}{\binom{N+1}{\frac{N}{2}+1}} \leqslant 1.$$

Therefore $S''' > T'''$.

As noted earlier, in other configurations the voter in a large state may not be more powerful than a voter in a small state.

With additional information specific cases can be analyzed. For example, it is possible to account for certain predispositions towards a candidate by allowing P to vary from voting unit to voting unit. The heuristic counterpart in the Shapley-Shubik formula is to have voters in some voting units be faster (or slower) in joining a coalition.[16]

The methodology presented here can be used to answer interesting constitutional questions. Thus power rankings in legislative, judicial and other voting bodies can be determined *a priori.*[17] Different constitutional arrangements can sometimes be shown to have similar power relationships (to use the terminology in the text, they are isomorphic).[18] While other constitutional arrangements which seem to be identical can be shown to have greatly differing power relationships (such as different size configurations in proportional weighted voting schemes). Thus the analysis presented here is useful not only for clarifying definitional problems but also for providing non obvious insight into a variety of phenomena.

3. Conclusion

The argument made here suggests that much of the theoretical and empirical work on power is much more general than previously believed. Results that seem to depend on a particular definition of power and a particular formula for calculating the probability of a coalition taking place have been shown to be equivalent to the results obtained when other definitions and formulae are used.

University of Chicago

NOTES

[1] William Riker, 'Some Ambiguities in the Notion of Power', *American Political Science Review* 58 (1964), 341-349. Excellent previous surveys not specifically mentioned elsewhere in my paper include Robert Dahl, 'Power', in *The International Encyclopedia of the Social Sciences,* Collier, 1966; James March, 'Power of Power', in David Easton (ed.), *Varieties of Political Theory,* Prentice Hall, New York, 1966, and Dorwin Cartwright, 'Influence, Leadership and Control', in James March (ed.), *Handbook of Organizations,* Rand McNally, Chicago. Roderick Bell, David Edwards and Harrison Wagner (eds.), *Political Power,* New York Free Press, 1969, is an excellent collection of articles on power including two important papers by J. Harsanyi as well as a number of articles mentioned elsewhere in this paper.

[2] For a discussion of power in games other than voting games see Donald Wittman, 'Various Concepts of Power', Equivalence Among Ostensibly Unrelated Approaches', *British Journal of Political Science* 6 (1977), 449-62.

[3] There are, of course, significant measurement problems remaining. For example, some wins may be more important than other wins. For a discussion of these problems see Duncan MacRae, Jr. and H. D. Price, 'Scale Positions and "Power" in the Senate', *Behavioral Science* (1959).

[4] Robert Dahl, James March, and D. Nasatir, 'Influence Ranking in the United State Senate', American Political Science Association Meetings (1956).

[5] Max Weber, *Theory of Social and Economic Organization,* edited by Talcott Parsons, Glencoe, Illinois, New Jersey Free Press, 1957, p. 152.

[6] James March, 'Measurement Concepts in the Theory of Influence', *Journal of Politics* **19** (1957), 202–226; J.W. Thibaut and H.H. Kelley, *The Social Psychology of Groups,* John Wiley and Sons, New York, 1959.

[7] George Karlsson, 'Some Aspects of Power in Small Groups', in Joan H. Caswell, Herbert Solomon and Patrick Suppes (eds.), *Mathematical Methods of Small Group Processes,* Stanford University Press, 1962, pp. 193–202.

[8] Jack Nagel, *The Descriptive Analysis of Power,* Yale University Press, New Haven, 1975. See also, Hayward Alker, 'On Political Capabilities in a Schedule Sense: Measuring Power, Integration and Development', in *Mathematical Approaches to Politics,* edited by Hayward Alker, Jr., Karl Deutsch and Antoine Stoetzee, Jossey-Bass, San Francisco, 1973.

[9] The relationship between player reversal and restricted outcome has already been noted by Dahl (1957) *op. cit.*, and March (1957) *op. cit.*, when there is no negative influence. There is negative influence if when a person is for, the probability of passage is less than when he is against. If this were true a person would just pretend he was against when he was for.

[10] L. S. Shapley, 'A Value for *N*-person Games', in H. W. Kuhn and A. W. Tucker (eds.), *Contributions to the Theory of Games II*, Annals of Mathematic Studies 28, Princeton University Press, 1953, New Jersey, pp. 307–317. L. S. Shapley and Martin Shubik, 'A Method for Evaluating the Distribution of Power in a Committee System', *The American Political Science Review* 48 (1954), 787–792. Irwin Mann and L. S. Shapley, 'The *A Priori* Voting Strength of Electoral College,' in Martin Shubik (ed.), *Game Theory and Related Approaches to Social Behavior,* John Wiley and Sons, New York, 1964, pp. 151–164. L. S. Shapley and William Riker, 'Weighted Voting: A Mathematical Analysis for Instrumental Judgements', J. Roland Pennock (ed.), *Representation: Nomox X*, New York, 1968, pp. 199–216.

[11] James Coleman, *The Mathematics of Collective Social Actions,* Aldine Atherton, Chicago, 1973. John Banzhaf, III, 'Weighted Voting Doesn't Work', *Rutgers Law Review* **XIX** (1965), 317–343.

[12] J. Coleman, 'The Marginal Utility of a Vote Commitment', *Public Choice* (1968).

[13] William Riker, *Theory of Political Coalitions,* Yale University Press, New Haven, 1962.

[14] In turn, this is isomorphic to the following situation. There are two groups each with one vote, two votes being needed for passage. In order for the one group composed of the five permanent members to vote for, all the permament members must vote for; in order for the group of six temporary members to vote for, two must vote for.

[15] The different approaches, however, will lead to different cardinal measures of power. See James Coleman, 'Control of Collectives and the Power of a Collectivity to Act', and Douglas Rae, 'Decisiveness in Election Outcomes', M. B. Leiberman (ed.), *Social Choice,* Gordon Breach, London, 1971. M. Allingham, 'Power and Value', University of Pennsylvania Fels Discussion Paper No. 43 (1973) and Fujihara, 'The Shapley-Shubik Power Index', University of Pennsylvania, unpublished manuscript, 1974.

[16] See Guillermo Owen, 'A Note on the Shapley Value', *Management Science* 14 (November 11, July, 1968), 731.

In more complicated games, it is possible for different voters to have differing numbers of votes even though the number of votes they have is irrelevant to whether the coalition is winning or not. For example, if there are three players, *A*, *B* and *C* with 9, 7 and 5 votes respectively, any two players can form a winning coalition. It is conceivable that the amount of apparent (rather than effective) resources (in this case votes) can affect the probabilities of various coalitions taking place. This view has been mainly expounded by sociologists Theodore Caplow, 'Theory of Coalition Formulation', *American Sociological Review* **21** (1956); William Gamson, 'A Theory of Coalition Formation', *American Sociological Review* **26** (1961), and Jerome Chertkoff, 'A Revision Caplow's Coalition Theory', *Journal of Experimental Social Psychology* **3** (1967), but is also an essential part of Riker's *Theory of Political Coalitions, op. cit.*, if minimal coalitions are not only viewed as being minimal in membership but also as being minimal in apparent resources. In the weighted voting game in the main body of the text, these issues do not come up since each voter has one vote.

Ideologies can also affect the probabilities of certain coalitions forming, for example, ideological conflicts can make it almost impossible for certain coalitions to take place. The analysis here is *a priori* and therefore ignores these ideological determinants of coalition formation. For an excellent collection of readings on coalition formation see Sven Gorennings, E. W. Kelley and Michael Leiserson, *The Study of Coalition Behavior,* Holt, Rinehart and Winston, New York. For more recent work, see Michael Taylor 'On the Theory of Government Coalition Formation', *British Journal of Politics* **2** (April, 1972), 361–386 and Abram DeSwaan *Coalition Theories and Cabinet Formations,* North-Holland, 1974.

[17] See Steven Brams and William Riker, 'Models of Coalition Formation in Voting Bodies', in James Herndon and Joseph Bernd (eds.), *Mathematical Applications in Political Science VI,* University of Virginia Press, Charlottesville, Virginia, 1972, who use both combinatorics and permutations to predict the dynamic aspects of coalition formation. For legislatures see Duncan Luce and Arnold Rogow, 'A Game Theoretic Analysis of Congressional Power Distribution for a State Two-party System', *Behavioral Science* **1** and **2** (April, 1956), 83–95; Samuel Krislov, 'Power and Coalition in a Nine-Man Body', *American Behavioral Scientist* **VI** (April, 1963), 24–26, and Glendon Schubert, 'The Power of Organized Minorities in Small Groups', *Administrative Science Quarterly* **IX** (September, 1964), 142–143, estimate power in the Supreme Court. Also see Chapter 6, 'The Power Index', in William Riker and Peter Ordeshook, *An Introduction to Positive Political Theory,* Prentice-Hall, New Jersey, 1973. A more recent application of the Shapley-Shubik formula has been made by Guillermo Owen 'Evaluation of a Presidential Election Game', *American Political Science Review* **LXIX** (September, 1975), 947–953. A recent survey is found in Steven Brams *Game Theory and Politics.*

[18] As another example not considered in the text: If we ignore the veto override, it makes no difference that the President cannot initiate legislation and can only accept or veto legislation passed by Congress. Since all three bodies must agree for legislation to be passed, it does not matter which initiates the legislation. This also holds when bargaining takes place. Actually this is a three stage voting game as the President and Congress are in turn elected by the voters. Type 3 games can be seen as coalition games. There is one coalition of size *N* and *L* individuals in coalitions by themselves. The coalition uses its own internal decision rule (the cases considered here have been majority rule).

INDEX OF NAMES

THE UNIVERSITY OF WESTERN ONTARIO SERIES IN PHILOSOPHY OF SCIENCE

A Series of Books on Philosophy of Science, Methodology, and Epistemology published in connection with the University of Western Ontario Philosophy of Science Programme

1. J. Leach, R. Butts, and G. Pearce (eds.), *Science, Decision and Value.* Proceedings of the Fifth University of Western Ontario Philosophy Colloquium, 1969. 1973, vii + 213 pp.
2. C. A. Hooker (ed.), *Contemporary Research in the Foundations and Philosophy of Quantum Theory.* Proceedings of a Conference held at the University of Western Ontario, London, Canada, 1973. xx + 385 pp.
3. J. Bub, *The Interpretation of Quantum Mechanics.* 1974, ix + 155 pp.
4. D. Hockney, W. Harper, and B. Freed (eds.), *Contemporary Research in Philosophical Logic and Linguistic Semantics.* Proceedings of a Conference held at the University of Western Ontario, London, Canada. 1975, vii + 332 pp.
5. C. A. Hooker (ed.), *The Logico-Algebraic Approach to Quantum Mechanics.* 1975, xv + 607 pp.
6. W. L. Harper and C. A. Hooker (eds.), *Foundations of Probability Theory, Statistical Inference, and Statistical Theories of Science,* 3 Volumes. Vol. I: *Foundations and Philosophy of Epistemic Applications of Probability Theory.* 1976, xi + 308 pp. Vol. II: *Foundations and Philosophy of Statistical Inference.* 1976, xi + 455 pp. Vol. III: *Foundations and Philosophy of Statistical Theories in the Physical Sciences.* 1976, xii + 241 pp.

8. J. M. Nicholas (ed.), *Images, Perception, and Knowledge.* Papers deriving from and related to the Philosophy of Science Workshop at Ontario, Canada, May 1974. 1977, ix+309 pp.
9. R. E. Butts and J. Hintikka (eds.), *Logic, Foundations of Mathematics, and Computability Theory.* Part One of the Proceedings of the Fifth International Congress of Logic, Methodology and Philosophy of Science, London, Ontario, Canada, 1975. 1977, x+406 pp.
10. R. E. Butts and J. Hintikka (eds.), *Foundational Problems in the Special Sciences.* Part Two of the Proceedings of the Fifth International Congress of Logic, Methodology and Philosophy of Science, London, Ontario, Canada, 1975. 1977, x+427 pp.
11. R. E. Butts and J. Hintikka (eds.), *Basic Problems in Methodology and Linguistics.* Part Three of the Proceedings of the Fifth International Congress of Logic, Methodology and Philosophy of Science, London, Ontario, Canada, 1975. 1977, x+321 pp.
12. R. E. Butts and J. Hintikka (eds.), *Historical and Philosophical Dimensions of Logic, Methodology and Philosophy of Science.* Part Four of the Proceedings of the Fifth International Congress of Logic, Methodology and Philosophy of Science, London, Ontario, Canada, 1975. 1977, x+336 pp.
13. C. A. Hooker (ed.), *Foundations and Applications of Decision Theory,* 2 volumes. Vol. I: *Theoretical Foundations.* 1978, xxiii+446 pp. Vol. II: *Epistemic and Social Applications.* 1978, xxiii+208 pp.
14. R. E. Butts and J. C. Pitt (eds.), *New Perspectives on Galileo.* Papers deriving from and related to a workshop on Galileo held at Virginia Polytechnic Institute and State University, 1975. 1978, xvi+262 pp.